Math Mammoth Grade 5 Tests and Cumulative Revisions

for the complete curriculum
(International Series)

Includes consumable student copies of:

- Chapter Tests
- End-of-year Test
- Cumulative Revisions
- Fraction Cut-outs

By Maria Miller

Contents

Grade 5, Chapter 1

End-of-Chapter Test

Instructions to the student:

Do not use a calculator. Answer each question in the space provided.

Instructions to the teacher:

My suggestion for grading the chapter 1 test is below. The total is 26 points. Divide the student's score by the total of 26 to get a decimal number, and change that decimal to percent to get the student's percentage score.

Question #	Max. points	Student score
1	3 points	
2	2 points	
3	2 points	
4	3 points	
5	2 points	

Question #	Max. points	Student score
6	2 points	
7	2 points	
8	4 points	
9	3 points	
10	3 points	
TOTAL	**26 points**	/ 26

Chapter 1 Test

1. Solve (without a calculator).

   a. $1456 \div 26$

   b. $18\,755 \div 31$

   c. 391×475

2. Solve: $Y - 8687 = 19\,764$

3. Solve in the right order.

a. $2 \times (80 - 8) =$ _____	**b.** $100 - 240 \div (8 + 2) =$ _____

4. Divide mentally.

a. $\dfrac{8109}{9} =$	**b.** $\dfrac{1244}{4} =$	**c.** $\dfrac{4045}{4+1} =$

5. Find a number to fit in the box so the equation is true.

a. $42 = (\; \boxed{} - 10\,) \times 2$	**b.** $48 \times 10 = \boxed{} \times 6$

6. Write an expression *or* an equation to match each written sentence. You do not have to solve anything.

a. The product of *s* and 11	**b.** The quotient of 48 and *b* is equal to 8.

7. Write a <u>single</u> expression (number sentence) for the problem, and solve.

> Mia bought 5 pairs of socks for $2.50 each, and paid with a $20 banknote. What was her change?

8. Draw a bar model to represent each equation and solve the equation.

a. $5 \times Y = 600$	**b.** $Z \div 3 = 140$

9. Is 991 divisible by 3?

Why or why not?

10. Factor the following numbers to their prime factors.

a. 16 /\	**b.** 34 /\	**c.** 80 /\

Grade 5, Chapter 2

End-of-Chapter Test

Instructions to the student:

Do not use a calculator. Answer each question in the space provided.

Instructions to the teacher:

My suggestion for grading the chapter 2 test is below. The total is 37 points. Divide the student's score by the total of 37 to get a decimal number, and change that decimal to percent to get the student's percentage score.

Question #	Max. points	Student score
1	3 points	
2	3 points	
3	8 points	
4	2 points	
5	3 points	

Question #	Max. points	Student score
6	8 points	
7	5 points	
8	4 points	
9	1 point	
TOTAL	**37 points**	/ 37

Chapter 2 Test

The calculator is not allowed for the first six problems of the test.

1. Write the numbers.

   a. 70 million 6 thousand 324

   b. 4 billion 32 thousand

   c. 98 billion 89 million 98

2. What is the *value* of the underlined digit?

a. 410 2<u>9</u>3 004	**b.** 408 0<u>3</u>7 443 000	**c.** <u>4</u> 395 490 493
Value: _____	Value: _____	Value: _____

3. Round these numbers as indicated.

number	183 602	355 079 933
to the nearest 1000		
to the nearest 10 000		
to the nearest 100 000		
to the nearest million		

4. Write using exponents, and solve.

   a. six squared =   **b.** two to the fifth power =

5. Solve.

a. $9^2 =$ _____	**b.** $10^3 =$ _____	**c.** $3^3 =$ _____

6. Solve.

a. $7 \times 10^4 =$ _____	**b.** $5604 \times 10\ 000 =$ _____
c. $10^7 \times 355 =$ _____	**d.** $10^5 \times 7900 =$ _____
e. $40 \times 900\ 000 =$ _____	**f.** $600 \times 200 \times 500 =$ _____
g. ▢ $\times 55 = 550\ 000\ 000$	**h.** _____ $\times 55\ 600 = 5\ 560\ 000\ 000$

7. Complete the maths path. (Calculator usage is optional.)

4 million	subtract 700 000 →	

add 12 million ↓

add 3 billion ↓

add 8 hundred ↓

subtract 20 thousand ←

8. First estimate using rounded numbers and mental maths. Then find the exact answer with a calculator.

a. What is your change, if you pay for five meals that cost $13.90 apiece with $100?	b. Find the total cost of purchasing 12 boxes of nails for $11.60 each.
My estimation: _____	My estimation: _____
_____	_____
Exact answer: _____	Exact answer: _____

9. A family purchases a new house for $324 400. They pay $40 000 of the price immediately and the rest in 12 equal payments. How much are those payments?

Grade 5, Chapter 3

End-of-Chapter Test

Instructions to the student:

Do not use a calculator. Answer each question in the space provided.

Instructions to the teacher:

My suggestion for grading the chapter 3 test is below. The total is 18 points. Divide the student's score by the total of 18 to get a decimal number, and change that decimal to percent to get the student's percentage score.

Question #	Max. points	Student score
1	4 points	
2	3 points	
3	3 points	

Question #	Max. points	Student score
4	3 points	
5	2 points	
6	3 points	
TOTAL	**18 points**	/ 18

Chapter 3 Test

1. First write the equation as the balance shows it. Then solve.

a.

X 47   X X

_____ = _____

_____ = _____

_____ = _____

b.

X X 47   X 51

_____ = _____

_____ = _____

_____ = _____

2. A cell phone that cost $96 is on sale with 1/6 off of the normal price.
How much would it cost for *three* discounted phones?

3. Matthew is 3/8 as tall as his dad.
If Matthew is 66 cm tall, then how tall is his dad?

4. Two sisters divided 250 smooth beach rocks so that
   the elder sister had 32 rocks more than the younger sister.
   How many rocks did the younger sister get?

5. Five kilograms of potatoes cost $7.50. Henry bought 2 kg.

   a. How much did 2 kg of potatoes cost?

   b. What was Henry's change from $10?

6. A high-quality hard drive costs three times as much
   as a low-quality one. Buying the two together would cost $820.
   How much does the low-quality hard drive cost?

Grade 5, Chapter 4

End-of-Chapter Test

Instructions to the student:

Do not use a calculator. Answer each question in the space provided.

Instructions to the teacher:

My suggestion for grading the chapter 4 test is below. The total is 52 points. Divide the student's score by the total of 52 to get a decimal number, and change that decimal to percent to get the student's percentage score.

Question #	Max. points	Student score
1	5 points	
2	2 points	
3	6 points	
4	4 points	
5	4 points	
6	4 points	

Question #	Max. points	Student score
7	6 points	
8	9 points	
9	2 points	
10	6 points	
11	4 points	
TOTAL	**52 points**	/ 52

Chapter 4 Test

1. Write the decimals indicated by the arrows.

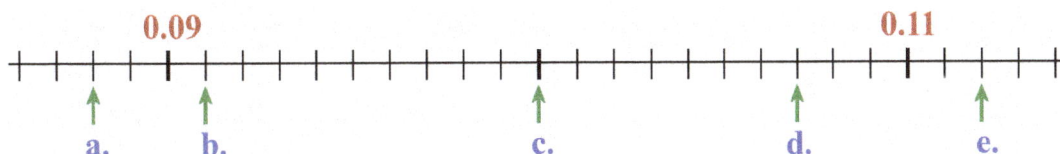

2. Write in expanded form.

   a. 0.253

   b. 5.07

3 Add and subtract.

a. $0.02 + 0.009 =$ _____	**b.** $1.3 - 0.7 =$ _____
c. $0.6 + 0.8 =$ _____	**d.** $0.9 - 0.02 =$ _____
e. $0.04 + 0.1 =$ _____	**f.** $2 - 0.12 =$ _____

4. Write as decimals.

a. $\dfrac{21}{100} =$	**b.** $\dfrac{46}{1000} =$	**c.** $3\dfrac{7}{100} =$	**d.** $20\dfrac{2}{10} =$

5. Write as fractions or mixed numbers.

   a. 0.6 **b.** 0.82 **c.** 1.208 **d.** 0.093

6. Compare.

   a. 0.05 ☐ 0.2 **b.** 0.43 ☐ 0.045 **c.** 2.05 ☐ 2.051 **d.** 0.438 ☐ $\dfrac{1}{2}$

7. Round the numbers to the nearest one, nearest tenth, and nearest hundredth.

rounded to...	nearest one	nearest tenth	nearest hundredth	rounded to...	nearest one	nearest tenth	nearest hundredth
8.816				0.398			
1.495				9.035			

8. Solve.

a. $0.4 \times 7 = $ _____	**b.** $20 \times 0.5 = $ _____	**c.** $0.02 \times 70 = $ _____
d. $7 \times 0.09 = $ _____	**e.** $10 \times 0.09 = $ _____	**f.** $0.8 \times 11 = $ _____
g. $0.24 \div 6 = $ _____	**h.** $0.081 \div 9 = $ _____	**i.** $5.6 \div 7 = $ _____

9. Find the number that is 1 tenth and 2 thousandths more than 1.109.

10. Add, subtract, and multiply.

a. $569 + 24.59 + 1.028$	**b.** $24.5 - 5.392$	**c.** 483×2.8
Estimate: _____	Estimate: _____	Estimate: _____

11. Divide using long division. Give your answers with two decimal digits.

a. $7.8 \div 5 = $ _____   **b.** $22 \div 3 = $ _____

Grade 5, Chapter 5

End-of-Chapter Test

Instructions to the student:

Do not use a calculator. Answer each question in the space provided.

Instructions to the teacher:

My suggestion for grading the chapter 5 test is below. The total is 17 points. Divide the student's score by the total of 17 to get a decimal number, and change that decimal to percent to get the student's percentage score.

Question #	Max. points	Student score
1	4 points	
2a	3 points	
2b	1 point	
3a	1 point	
3b	2 points	
3c	1 point	

Question #	Max. points	Student score
4a	2 points	
4b	1 point	
4c	1 point	
4d	1 point	
TOTAL	**17 points**	/ 17

Chapter 5 Test

1. Plot the points from the "number rule" on the coordinate grid.

   x-values: start at 1, and add 1 each time.

   y-values: start at 1, and add 2 each time.

x	1	2	3	4
y				

x	5	6	7	8
y				

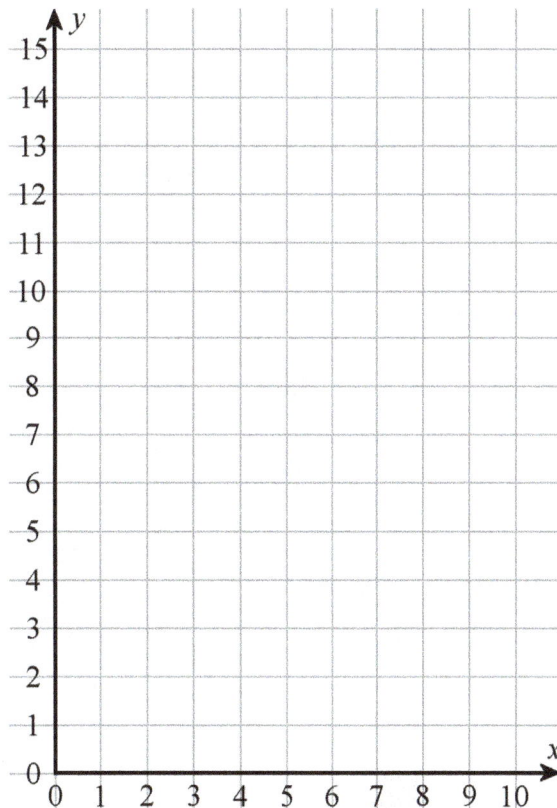

2. **a.** Plot the points A, B, and C on the grid. Note that the points don't necessarily fall on the gridlines.

   A(2, 3)   B(2 , 13)   C(7 ½ , 13)

   b. Plot point D in such a manner that when you connect A, B, C, and D with line segments, ABCD forms a rectangle.

   What are the coordinates of D?

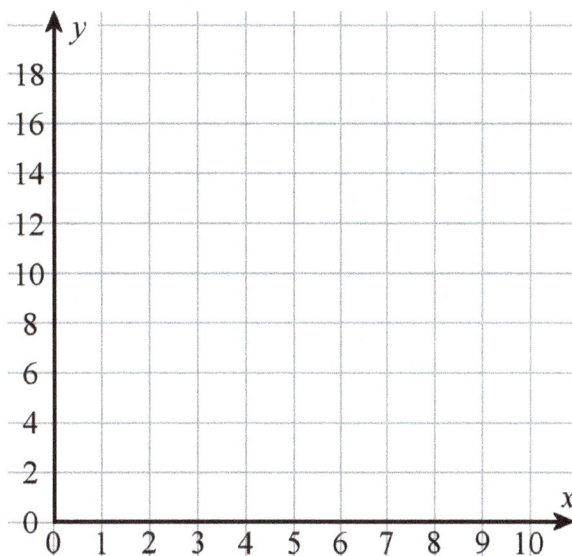

3. A store kept track of how many cell phones they sold each day of the week ("units sold").
On a certain day they started a promotion with 3/10 off of the normal price.

a. Add the number labels for the vertical axis next to the tick marks (the scaling).

b. Plot the remaining points and finish the line graph.

c. Which day did the promotion most likely start?

Day	Units sold
Mo	17
Tu	14
Wd	15
Th	21
Fr	19
Sa	23
Mo	15
Tu	34
Wd	40
Th	37
Fr	33
Sa	41

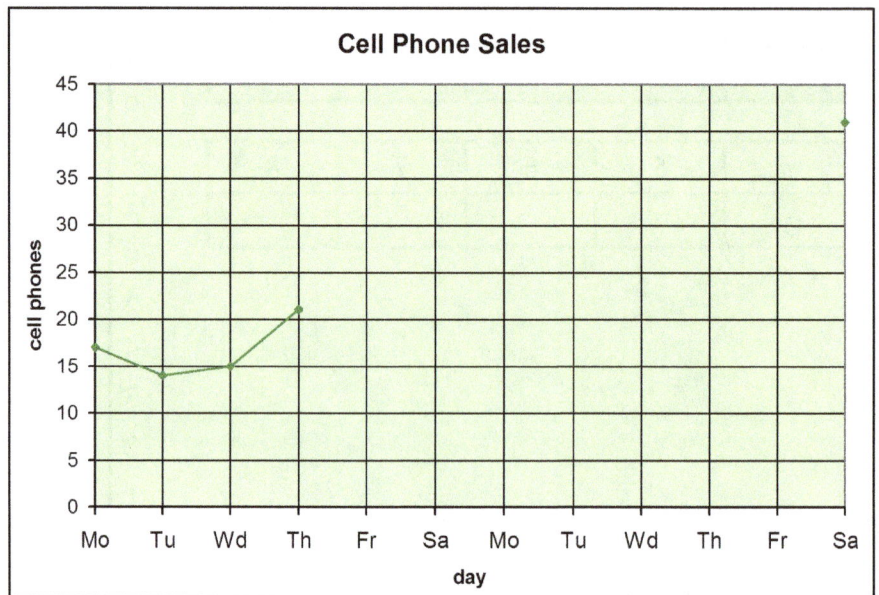

4. The chart shows Alice's science test scores for five different tests.

Alice's test scores	
Test 1	76
Test 2	66
Test 3	74
Test 4	81
Test 5	88

a. Draw a line graph.

b. Calculate the average.

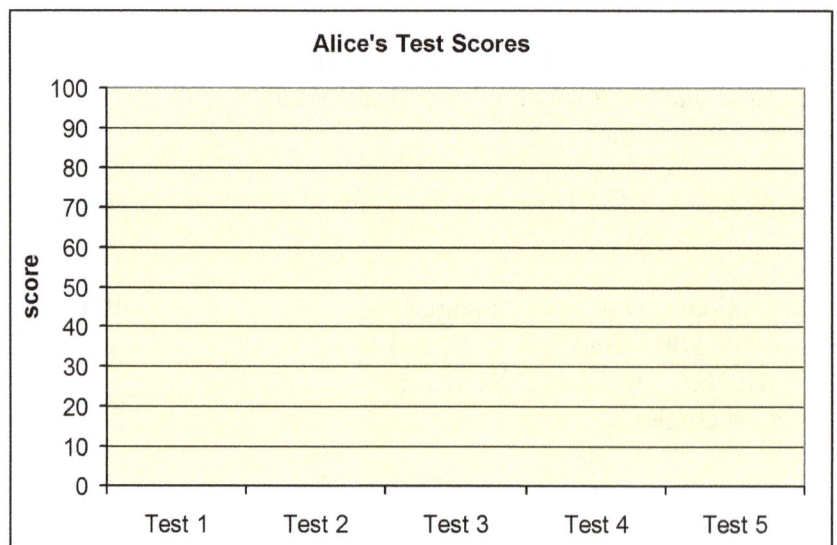

c. Plot the average on the line graph.

d. In which tests did Alice score below her average?

Grade 5, Chapter 6

End-of-Chapter Test

Instructions to the student:

Do not use a calculator. Answer each question in the space provided.

Instructions to the teacher:

My suggestion for grading the chapter 6 test is below. The total is 59 points. Divide the student's score by the total of 59 to get a decimal number, and change that decimal to percent to get the student's percentage score.

Question #	Max. points	Student score
1	6 points	
2	6 points	
3	4 points	
4	2 points	
5	3 points	
6	2 points	
7	2 points	
8	6 points	

Question #	Max. points	Student score
9	6 points	
10	2 points	
11	4 points	
12	4 points	
13	4 points	
14	4 points	
15	4 points	
TOTAL	**59 points**	/ 59

Chapter 6 Test

1. Solve.

a. $0.4 \times 0.07 = $ _____	**c.** $0.12 \times 0.5 = $ _____	**e.** $0.2 \times 1000 = $ _____
b. $7 \times 0.09 = $ _____	**d.** $100 \times 0.09 = $ _____	**f.** $1.1 \times 0.6 = $ _____

2. Divide.

a. $0.24 \div 0.04 = $ _____	**c.** $2 \div 100 = $ _____	**e.** $0.43 \div 10 = $ _____
b. $0.045 \div 0.009 = $ _____	**d.** $0.8 \div 10 = $ _____	**f.** $7 \div 1000 = $ _____

3. Multiply and divide.

a. $0.05 \times 10^4 = $ _____	**c.** $3.5 \div 10^2 = $ _____
b. $10^5 \times 7.8 = $ _____	**d.** $13\ 200 \div 10^4 = $ _____

4. **a.** Estimate the answer to 0.6×21.8.

   b. Now find the exact answer to 0.6×21.8.

5. Teresa packed 7 kg of blueberries equally into four boxes. How much does each box weigh?

6. Find 3/10 of 225 kg.

7. Is the answer to 0.9 × 0.8 more or less than 0.8?

Explain in your own words why that is so.

8. Convert.

a. 0.7 m = _____ cm	**b.** 2650 ml = _____ L	**c.** 5.16 kg = _____ g
3.2 km = _____ m	0.9 L = _____ ml	400 g = _____ kg

9. Convert.

a. 8 kg 10 g = _____ g	**b.** 2 km 3 m = _____ m	**c.** 81 cm = _____ m
183 cm = _____ m _____ cm	45 L = _____ ml	165 g = _____ kg

10. Brett has $1.45 in his pocket — and it's all in nickels (5-cent coins)!
    Write a division <u>by a decimal</u> that will solve how many nickels he has.

11. Samuel bought a 0.9-litre box of juice and two cans of juice, 350 ml each.
    What is the total volume of the juice he bought?

12. Mary bought a 2-kg bag of turnips for $4.48.
    Then, Mary sold 250 g of the turnips to her friend.

    a. How much does half a kilogram of turnips cost?

    b. How much did she charge her friend?

13. Divide using long division. If necessary, round your answer to two decimal digits.

   a. 7.8 ÷ 0.005 = _____

   b. 4.39 ÷ 0.3 = _____

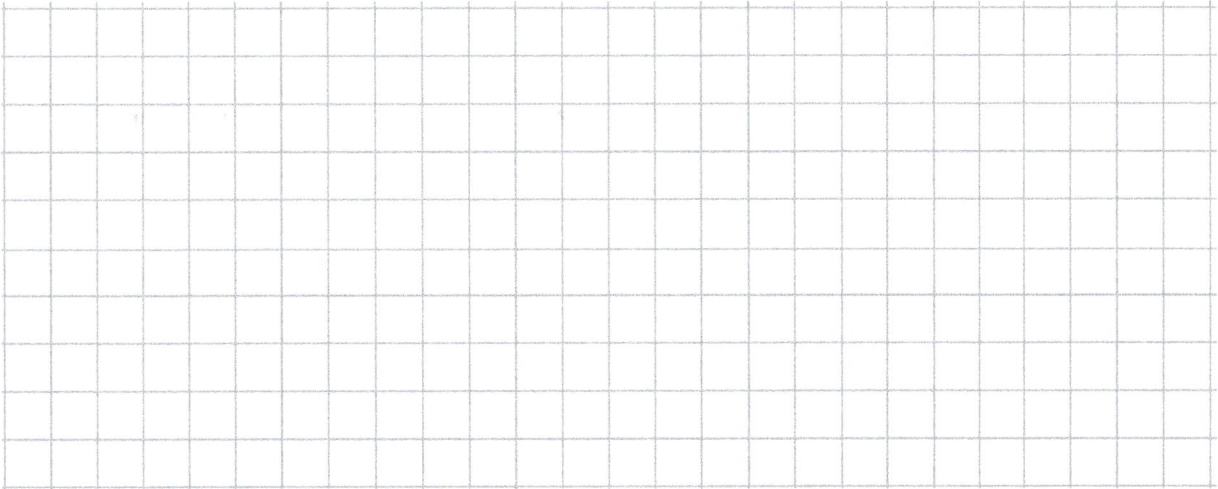

14. A keyboard that normally costs $19.95 is discounted.
    The new price is 2/5 off of the normal price.
    Find how much *two* discounted keyboards cost.

15. Arnold weighed the apples he had left in the root cellar in the spring. He gave half of the
    apples to his neighbour, and divided the rest equally into four boxes. Each box weighed 0.47 kg.

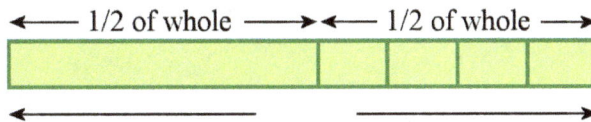

    ←——— 1/2 of whole ———→ ←——— 1/2 of whole ———→

    a. In the model, label the parts that equal 0.47 kg.

    b. In the model, label the total weight of all the apples with *x*.

    c. Find the total weight of all the apples.

Grade 5, Chapter 7

End-of-Chapter Test

Instructions to the student:

Do not use a calculator. Answer each question in the space provided.

Instructions to the teacher:

If you are using the digital version, the test is best printed at 100%, not at "shrink to fit", or "print to fit", or similar setting in your printer. This is because question #9 has a triangle for the student to measure. If you don't print it at 100%, the measurements won't match the answer key. It is not a big problem — simply check the student's measurements.

My suggestion for grading the chapter 7 test is below. The total is 39 points. Divide the student's score by the total of 39 to get a decimal number, and change that decimal to percent to get the student's percentage score.

Question #	Max. points	Student score
1	3 points	
2	3 points	
3	5 points	
4	5 points	
5	5 points	
6	3 points	
7	4 points	

Question #	Max. points	Student score
8	2 points	
9	3 points	
10a	1 point	
10b	1 point	
10c	1 point	
10d	3 points	
TOTAL	**39 points**	/ 39

Chapter 7 Test

1. Write as mixed numbers.

     a. $\dfrac{26}{3}$ **b.** $\dfrac{45}{7}$ **c.** $\dfrac{34}{5}$

2. Add or subtract.

a. $\quad 7\dfrac{6}{8}$ $\quad +\; 2\dfrac{5}{8}$	**b.** $\quad 6\dfrac{1}{5}$ $\quad -\; 3\dfrac{4}{5}$	**c.** $\quad 4\dfrac{6}{11}$ $\quad 9\dfrac{9}{11}$ $\quad +\; 2\dfrac{4}{11}$

3. Mark the fractions on the number line. $\dfrac{2}{3},\ \dfrac{5}{6},\ \dfrac{7}{12},\ \dfrac{3}{4},\ \dfrac{11}{12}$

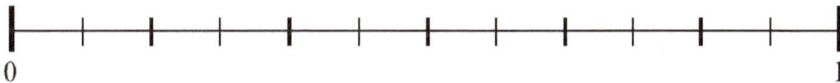

         0   1

4. If you can find an equivalent fraction, write it. If you cannot, cross out the whole problem.

a. $\dfrac{3}{7} = \dfrac{}{21}$	**b.** $\dfrac{4}{3} = \dfrac{}{18}$	**c.** $\dfrac{5}{6} = \dfrac{}{11}$	**d.** $\dfrac{2}{5} = \dfrac{8}{}$	**e.** $\dfrac{5}{6} = \dfrac{15}{}$

5. Compare the fractions, and write < , >, or = in the box.

a. $\dfrac{7}{4}\ \square\ \dfrac{5}{3}$	**b.** $\dfrac{5}{11}\ \square\ \dfrac{1}{2}$	**c.** $\dfrac{7}{10}\ \square\ \dfrac{69}{100}$	**d.** $\dfrac{3}{4}\ \square\ \dfrac{75}{100}$	**e.** $\dfrac{8}{7}\ \square\ \dfrac{7}{9}$

6. Draw something in the picture and explain how we can add 1/3 and 2/5.

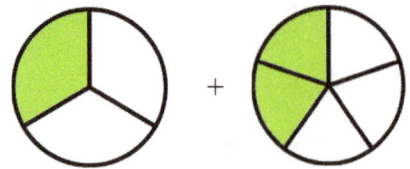

7. Add and subtract.

a. $\dfrac{2}{3} + \dfrac{3}{4}$	**b.** $\dfrac{5}{6} - \dfrac{2}{3}$
c. $3\dfrac{1}{7} - \dfrac{1}{2}$	**d.** $6\dfrac{7}{8} + 3\dfrac{1}{5}$

8. Write the fractions in order starting from the smallest.

$$\dfrac{4}{7}, \ \dfrac{5}{9}, \ \dfrac{7}{5}, \ \dfrac{1}{2}$$

9. During a snowy week, Alice kept track of how much of their family's driveway she shovelled clear of snow (other family members did the rest).

On Monday, she shovelled half of it. On Tuesday and Wednesday, one-third of it, on Thursday a quarter of it, and on Friday 3/8 of it. How many times did she shovel the equivalent of the entire driveway, in total?

10. Shaun has a job of delivering newspapers with advertisements every Wednesday. Over 10 weeks, he kept track of how long it took him to deliver them on his route.

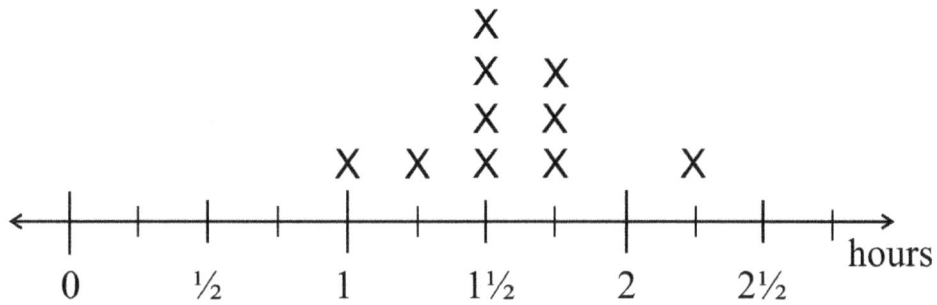

```
                        X
                        X   X
                        X   X
            X   X   X   X       X
   +---+---+---+---+---+---+---+---+---+---+--->
   0      ½       1      1½      2      2½     hours
```

a. How many times (out of these 10 weeks) did it take him 1 hour 45 minutes to do so?

b. How long did it take him when it took him the longest (it was bad weather)?

c. What is the most common amount of time it took him?

d. How much time in total did he spend delivering newspapers in these 10 weeks?

Grade 5, Chapter 8

End-of-Chapter Test

Instructions to the student:

Do not use a calculator. Answer each question in the space provided.

Instructions to the teacher:

My suggestion for grading the chapter 8 test is below. The total is 32 points. Divide the student's score by the total of 32 to get a decimal number, and change that decimal to percent to get the student's percentage score.

Question #	Max. points	Student score
1	3 points	
2	2 points	
3	3 points	
4	4 points	
5	6 points	
6	4 points	

Question #	Max. points	Student score
7	2 points	
8	2 points	
9	1 point	
10	2 points	
11	3 points	
TOTAL	**32 points**	/ 32

Chapter 8 Test

1. If possible, simplify the following fractions. Give your answer as a mixed number when possible.

a. $\dfrac{22}{6} =$	**b.** $\dfrac{28}{42} =$	**c.** $\dfrac{35}{32} =$

2. Julie needs 2/3 cup of butter for one batch of cookies.
   Find how much butter she would need to make five batches of cookies.

3. Is the following multiplication correct?
   If not, correct it.

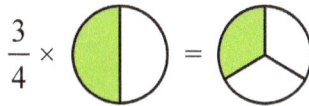

   $$\dfrac{3}{4} \times \bigcirc = \bigcirc$$

4. Is the result of multiplication more, less, or equal to the original number? You do not have to calculate anything. Compare, writing $<$, $>$, or $=$ in the box.

a. $\dfrac{23}{20} \times 73 \; \square \; 73$	**b.** $\dfrac{49}{49} \times 1.45 \; \square \; 1.45$	**c.** $\dfrac{4}{9} \times \dfrac{7}{6} \; \square \; \dfrac{7}{6}$	**d.** $\dfrac{9}{11} \times 795 \; \square \; 795$

5. Multiply the fractions, and shade a picture to illustrate the multiplication. Simplify your answers.

a. $\dfrac{2}{3} \times \dfrac{1}{6}$	**b.** $\dfrac{4}{9} \times \dfrac{2}{3}$

6. Multiply. Give your answers in the lowest terms (simplified) and as a mixed number, if possible.

a. $\dfrac{5}{12} \times \dfrac{2}{3}$	**b.** $9 \times \dfrac{4}{5}$

7. Find the area of a square with 1 3/8-unit sides.

8. After supper, a family of four had 1/3 of a pizza left.
   The next day, three people shared the remaining pizza equally.
   What fractional part of the *whole* pizza did each person get?

9. Five granola bars were divided equally between 8 people.
   How much of a bar did each person get?

10. **a.** How many 1/3-kg bags can you get from 3 kg of cocoa?

    b. Write a division sentence to match this situation.

11. Solve.

a. $\dfrac{1}{6} \div 3$	**b.** $6 \div \dfrac{1}{8}$	**c.** $\dfrac{9}{11} \div 3$
d. $10 \div \dfrac{1}{3}$	**e.** $\dfrac{8}{15} \div 4$	**f.** $\dfrac{1}{2} \div 7$

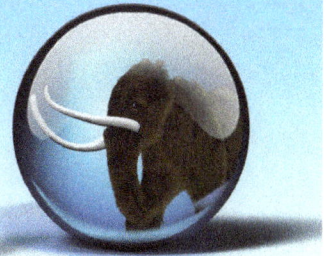

Grade 5, Chapter 9

End-of-Chapter Test

Instructions to the student:

Do not use a calculator. Answer each question in the space provided.

Instructions to the teacher:

My suggestion for grading the chapter 9 test is below. The total is 34 points. Divide the student's score by the total of 34 to get a decimal number, and change that decimal to percent to get the student's percentage score.

Question #	Max. points	Student score
1	2 points	
2	3 points	
3a	1 point	
3b	1 point	
3c	2 points	
3d	1 point	
4	3 points	
5	3 points	
6	2 points	
7	2 points	

Question #	Max. points	Student score
8a	2 points	
8b	2 points	
9	2 points	
10a	1 point	
10b	2 points	
11	2 points	
12	1 point	
13a	1 point	
13b	1 point	
TOTAL	**34 points**	/ 34

Chapter 9 Test

1. What is the name of a polygon...

   a. with eight sides?

   b. with six sides that are all the same length?

2. Plot the points (0, 2), (0, 7), (5, 4), and connect them with line segments to form a triangle.

   Classify the triangle by its angles and sides. The triangle is

   _____ and

   _____

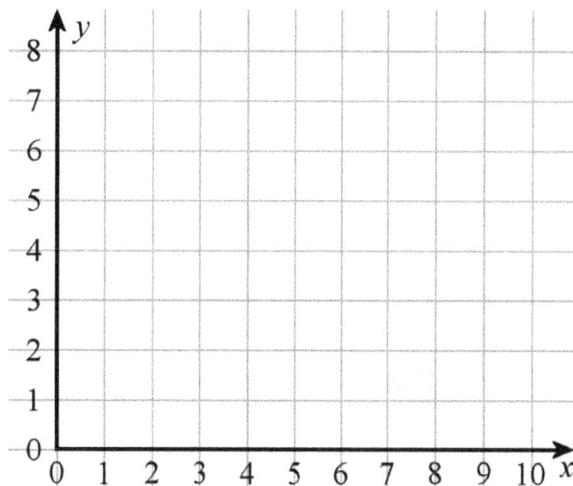

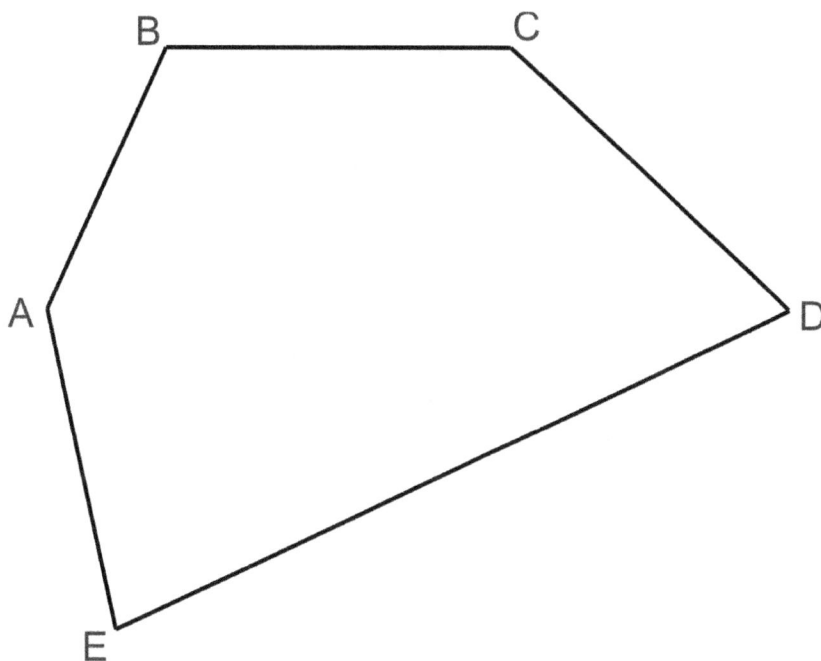

3. **a.** What is shape ABCDE called?

   b. Draw a diagonal from A to D.

   c. What two shapes are now formed? Use the most descriptive names.

   d. Measure angle D.

4. Name the shapes. Use the most descriptive names.

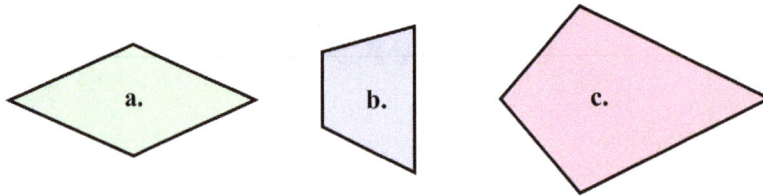

a. _____   b. _____

c. _____

5. Classify each triangle according to its sides and its angles.

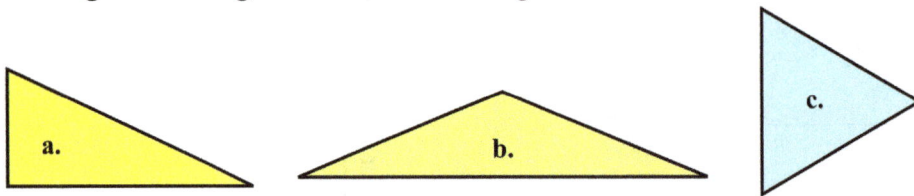

a. _____ and _____

b. _____ and _____

c. _____ and _____

6. A certain quadrilateral has two pairs of parallel sides, and
   its sides measure 5 cm, 5 cm, 5 cm, and 5 cm.
   What type of quadrilateral is it? Sketch example(s).

7. A quadrilateral has exactly one pair of parallel sides, and
   exactly one pair of congruent sides. What type of quadrilateral
   is it? Sketch example(s).

8. Answer.

   a. Could a scalene triangle be an obtuse triangle? (Yes / No)
      If yes, sketch an example.

   b. Could an acute triangle be isosceles? (Yes / No)
      If yes, sketch an example.

9. A square has a perimeter of 4 units. What is its area?

10. The edge of each little cube in the shape on the right measures 2 cm.

   a. What is the volume of one little cube?

   b. What is the total volume of the shape, in cubic centimetres?

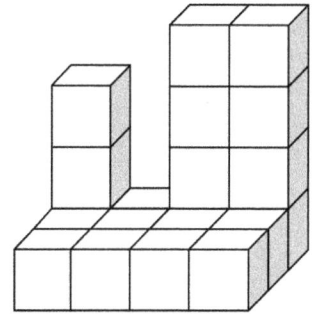

11. The dimensions of this box are 2 units by 1.5 units by 1.5 units. What is its volume?

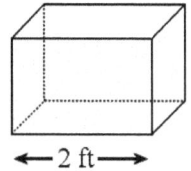

←—2 ft—→

12. The volume of a rectangular prism is 30 m^3, and its bottom area measures 3 m by 2 m. How tall is the prism?

13. A book measures 15 cm × 30 cm × 1.5 cm. You make a stack of six books.

   a. What is the volume of one book?

   b. What is the volume of the stack?

End-of-Year Test - Grade 5
International Version

This test is quite long, because it contains lots of questions on all of the major topics covered in *Math Mammoth Grade 5 International Version*. Its main purpose is to be a diagnostic test—to find out what the student knows and does not know. The questions are quite basic and do not involve especially difficult word problems.

Since the test is so long, I do not recommend that you have your child/student do it in one sitting. Break it into 3-5 parts and administer them on consecutive days, or perhaps on morning/evening/morning/evening. Use your judgement.

A calculator is not allowed.

The test is evaluating the student's ability in the following content areas:

- the four operations with whole numbers
- the concept of an equation; solving simple equations
- divisibility and factoring
- place value and rounding with large numbers
- solving word problems, especially those that involve a fractional part of a quantity
- the concept of a decimal and decimal place value
- all four operations with decimals, to the hundredths
- coordinate grid, drawing a line graph, and finding the average
- fraction addition and subtraction
- equivalent fractions and simplifying fractions
- fraction multiplication
- division of fractions in special cases (a unit fraction divided by a whole number, and a whole number divided by a unit fraction)
- classifying triangles and quadrilaterals
- volume of rectangular prisms (boxes)

In order to continue with the *Math Mammoth Grade 6*, I recommend that the child gain a minimum score of 80% on this test, and that the teacher or parent revise with them any content areas in which the child may be weak. Children scoring between 70% and 80% may also continue with grade 6, depending on the types of errors (careless errors or not remembering something, versus a lack of understanding). Again, use your judgement.

Instructions to the student:

Do not use a calculator. Answer each question in the space provided.

Instructions to the teacher: The total is 182 points. A score of 146 points is 80%.

Question #	Max. points	Student score
The Four Operations		
1	2 points	
2	6 points	
3	2 points	
4	4 points	
5	2 points	
6	2 points	
7	3 points	
	subtotal	/ 21
Large Numbers		
8	2 points	
9	1 point	
10	4 points	
11	1 point	
12	4 points	
	subtotal	/ 12
Problem Solving		
13a	2 points	
13b	2 points	
14	3 points	
15	3 points	
16	3 points	
17	3 points	
	subtotal	/ 16
Decimals		
18	4 points	
19	6 points	
20	3 points	
21	3 points	
22	3 points	
23	3 points	
24	9 points	
25	6 points	
26	9 points	
27	3 points	
28	3 points	
	subtotal	/52

Question #	Max. points	Student score
Graphs		
29	3 points	
30	4 points	
	subtotal	/7
Fractions		
31	3 points	
32	4 points	
33	4 points	
34	2 points	
35	4 points	
36	2 points	
37	5 points	
38	3 points	
39	2 points	
40	4 points	
41	2 points	
42	2 points	
43	4 points	
44	4 points	
	subtotal	/45
Geometry		
45	4 points	
46	3 points	
47	6 points	
48	2 points	
49	2 points	
50	3 points	
51a	1 point	
51b	2 points	
52	2 points	
53	4 points	
	subtotal	/29
	TOTAL	/182

48

Math Mammoth End-of-Year Test - Grade 5

The Four Operations

1. Solve (without a calculator).

   a. $1035 \div 23$

   b. 492×832

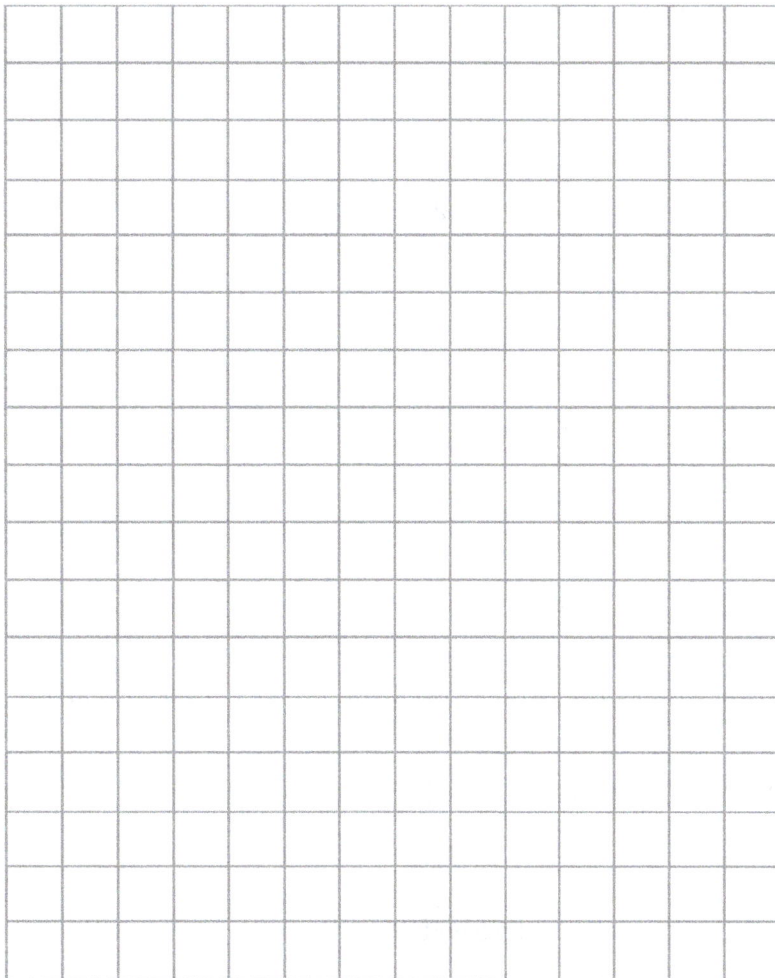

2. Solve.

   a. $x - 56\,409 = 240\,021$

   b. $7200 \div Y = 90$

   c. $N \div 14 = 236$

3. Write an equation to match this model, and solve it.

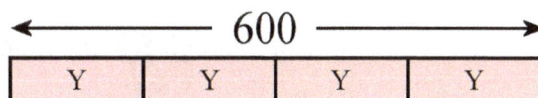

   $\longleftarrow$ 600 $\longrightarrow$

Y	Y	Y	Y

4. Solve in the right order.

a. $(25 + 8) \div 3 \times 2 = \underline{\hspace{2cm}}$	**b.** $2 \times (30 - 12 + 2) + 8 = \underline{\hspace{2cm}}$
c. $25 + 8 \times 3 \div 2 = \underline{\hspace{2cm}}$	**d.** $10 \div 2 \times (7 + 8) = \underline{\hspace{2cm}}$

5. Place parentheses into the equations to make them true.

    a. $42 \times 10 = 10 - 4 \times 70$ **b.** $143 = 13 \times 5 + 6$

6. Is 991 divisible by 4?

   Why or why not?

7. Factor the following numbers to their prime factors.

a. 26 / \	**b.** 40 / \	**c.** 59 / \

Large Numbers

8. Write the numbers.

    a. 70 million 16 thousand 90

    b. 32 billion 232 thousand

9. What is the value of the digit 8 in the number **56 782 010 000**?

10. Calculate the products.

a. 224×10^7	**b.** $78\,009 \times 10^5$
c. $30\,000 \times 5000$	**d.** $400 \times 20 \times 60$

11. Estimate the result of 31 933 × 305.

12. Round these numbers to the nearest thousand, nearest ten thousand, nearest hundred thousand, and nearest million.

number	593 204	19 054 947
to the nearest 1000		
to the nearest 10 000		
to the nearest 100 000		
to the nearest million		

Problem Solving

13. Write a single expression (number sentence) for each problem, and solve.

a. A store was selling movies that originally cost $19.95 with a $5 discount. Mia bought five of them. What was the total cost?

Expression: _____

b. A website charges a fixed amount for each song download. If you can download six songs for $4.68, then how much would it cost to download ten songs?

Expression: _____

14. Jack has an 3-metre long board. He cuts off 1/6 of it.
    How long is the remaining piece, in metres and centimetres?

15. A blue swimsuit costs $42 and a red swimsuit
    costs 5/6 as much. How much would the two
    swimsuits cost together?

    Mark the $42 in the bar model. Mark what is not
    known with "?". Solve.

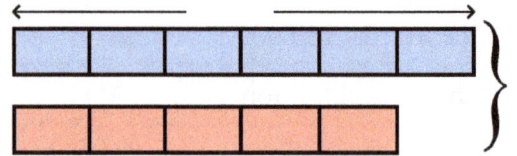

16. A bag has green and purple marbles. Two-fifths of the marbles are green, and the rest are purple.

    a. Draw a bar model for this situation.

    b. If there are 134 green marbles, how many are purple?

17. Karen and Ann share the cost of a DVD that costs $29.90
    so that Karen pays 3/5 of it and Ann pays 2/5 of it.

    a. *Estimate* how much each person will pay.

    b. Find the exact amount of how much each person will pay.

Decimals

18. Write the decimals indicated by the arrows.

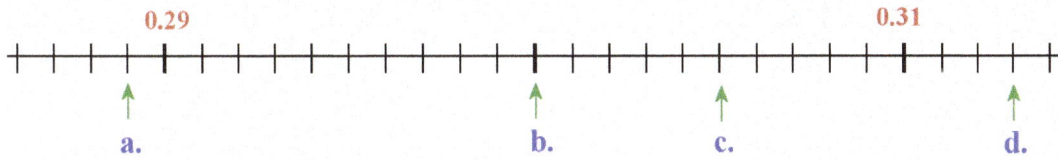

a. _____ b. _____ c. _____ d. _____

19. Complete.

a. $0.9 + 0.05 =$ _____	**b.** $0.28 +$ _____ $= 1$	**c.** $0.82 - 0.2 =$ _____
d. $1.3 - 0.04 =$ _____	**e.** $0.25 + 0.8 =$ _____	**f.** _____ $- 0.2 = 0.17$

20. Write as decimals.

   a. $\dfrac{8}{100} =$
     b. $\dfrac{81}{1000} =$
     c. $5\dfrac{21}{100} =$

21. Write as fractions or mixed numbers.

   a. 0.048
     b. 1.004
     c. 7.22

22. Compare, and write $<$ or $>$.

   a. 0.31 ☐ 0.031
   b. 0.43 ☐ 0.093
   c. 1.6 ☐ 1.29

23. Round the numbers to the nearest one, nearest tenth, and nearest hundredth.

rounded to...	nearest one	nearest tenth	nearest hundredth	rounded to...	nearest one	nearest tenth	nearest hundredth
5.098				0.306			

24. Solve.

a. $0.4 \times 7 =$	**d.** $10 \times 0.05 =$	**g.** $1.1 \times 0.3 =$
b. $0.4 \times 0.7 =$	**e.** $1000 \times 0.05 =$	**h.** $70 \times 0.9 =$
c. $0.4 \times 700 =$	**f.** $10^5 \times 0.5 =$	**i.** $20 \times 0.09 =$

25. Divide.

a. $0.36 \div 6 =$	**c.** $3 \div 100 =$	**e.** $16 \div 10^2 =$
b. $5.6 \div 7 =$	**d.** $0.7 \div 10 =$	**f.** $712 \div 10^3 =$

26. Convert.

a. 0.2 m = _____ cm	**b.** 0.4 L = _____ ml	**c.** 3670 mm = _____ m _____ mm
37 cm = _____ m	3.5 kg = _____ g	465 cm = _____ m _____ cm
2.9 km = _____ m	240 g = _____ kg	4060 g = _____ kg _____ g

27. Two litres of ice cream is divided equally into nine bowls. Calculate, to the nearest millilitre, how much ice cream is in *two* bowls.

28. Calculate.

    a. $4.2 - 2.78$

    b. $71.40 \div 5$

    c. 2.2×6.4

Graphs

29. Plot the points from the "number rule" on the coordinate grid.

The rule for *x*-values:
Start at 0, and add 1 each time.

The rule for *y*-values:
Start at 1, and add 2 each time.

x	0	1				
y	1					

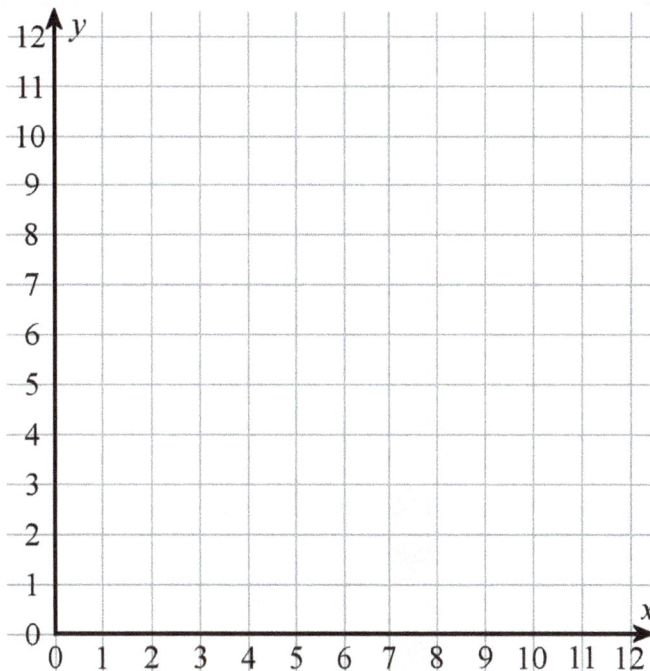

30. The table below gives the amount of sales in a grocery store from Monday through Friday.

Day	Sales (thousands of dollars)
Mon	125
Tue	114
Wed	118
Thu	130
Fri	158

a. Make a line graph.

b. Calculate the average daily sales for this period.

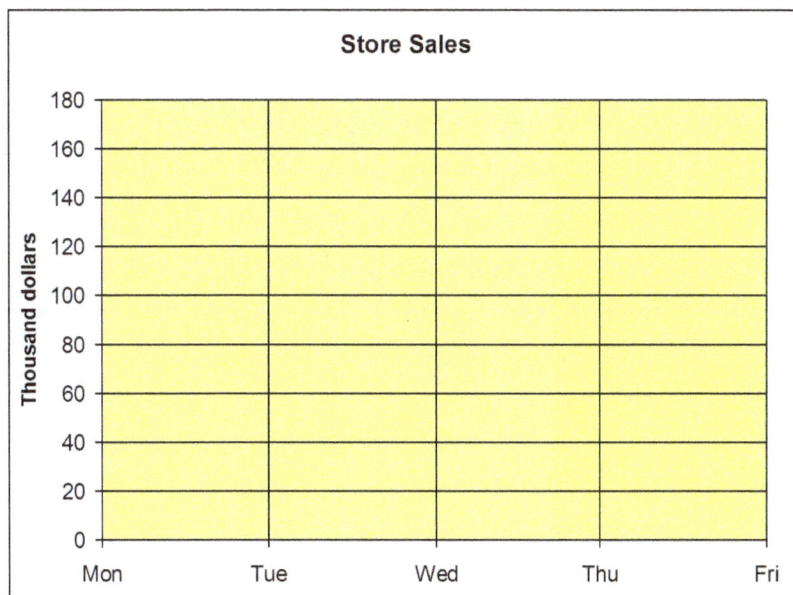

Fractions

31. Add and subtract.

a.	b.	c.
$3\dfrac{7}{9}$ $+\ \ 2\dfrac{5}{9}$ ——————	$5\dfrac{1}{6}$ $-\ \ 2\dfrac{5}{6}$ ——————	$3\dfrac{7}{10}$ $2\dfrac{8}{10}$ $+\ 7\dfrac{3}{10}$ ——————

32. Mark the fractions on the number line. $\dfrac{3}{4}$, $\dfrac{1}{3}$, $\dfrac{4}{6}$, $\dfrac{5}{12}$

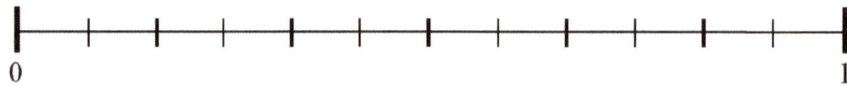

33. If you can find an equivalent fraction, write it. If you cannot, cross the whole problem out.

a. $\dfrac{5}{6} = \dfrac{\ }{20}$	b. $\dfrac{2}{7} = \dfrac{\ }{28}$	c. $\dfrac{3}{8} = \dfrac{15}{\ }$	d. $\dfrac{2}{9} = \dfrac{6}{\ }$

34. Find the errors in Mia's calculation and correct them.

"I need these to have the same denominator."

$$\dfrac{2}{5} + \dfrac{2}{3}$$

$$\downarrow \qquad \downarrow$$

$$\dfrac{2}{15} + \dfrac{2}{15} = \dfrac{4}{15}$$

35. Add and subtract the fractions and mixed numbers.

a. $\dfrac{1}{3} + \dfrac{5}{6}$	**b.** $\dfrac{4}{5} - \dfrac{1}{3}$
c. $6\dfrac{1}{8} - \dfrac{1}{2}$	**d.** $6\dfrac{7}{9} + 3\dfrac{1}{2}$

36. You need 2 3/4 cups of flour for one batch of rolls.
Find how much flour you would need for three batches of rolls.

37. Compare the fractions, and write $<$, $>$, or $=$ in the box.

a. $\dfrac{6}{9} \,\square\, \dfrac{6}{13}$
b. $\dfrac{6}{13} \,\square\, \dfrac{1}{2}$
c. $\dfrac{5}{10} \,\square\, \dfrac{48}{100}$
d. $\dfrac{1}{4} \,\square\, \dfrac{25}{100}$
e. $\dfrac{5}{7} \,\square\, \dfrac{7}{10}$

38. Simplify the following fractions if possible. Give your answer as a mixed number when you can.

a. $\dfrac{21}{15} =$	**b.** $\dfrac{29}{36} =$	**c.** $\dfrac{42}{48} =$

39. Is the following multiplication correct?
If not, correct it.

$\dfrac{2}{3} \times$ $=$

40. Multiply the fractions, and shade a picture to illustrate the multiplication.

a. $\dfrac{1}{3} \times \dfrac{5}{6}$ **b.** $\dfrac{2}{9} \times \dfrac{2}{3}$

41. How many 1/4-metre pieces can you cut from a string that is 15 m long?

42. Three people share half a pizza evenly. What fractional part of the original pizza does each one get?

43. Is the result of multiplication more, less, or equal to the original number? You do not have to calculate anything. Compare, writing $<$, $>$, or $=$ in the box.

| **a.** $\dfrac{19}{17} \times 93 \;\square\; 93$ | **b.** $\dfrac{8}{9} \times \dfrac{5}{6} \;\square\; \dfrac{5}{6}$ | **c.** $\dfrac{14}{15} \times 516 \;\square\; 516$ | **d.** $\dfrac{52}{52} \times 7.09 \;\square\; 7.09$ |

44. Solve. Give your answer as a mixed number and simplified to lowest terms.

a. $\dfrac{7}{6} \times 9$	**b.** $\dfrac{1}{7} \div 3$
c. $\dfrac{4}{5} \times 3\dfrac{2}{3}$	**d.** $2 \div \dfrac{1}{9}$

58

Geometry

45. Measure the sides of the triangle in centimetres. Find its perimeter.

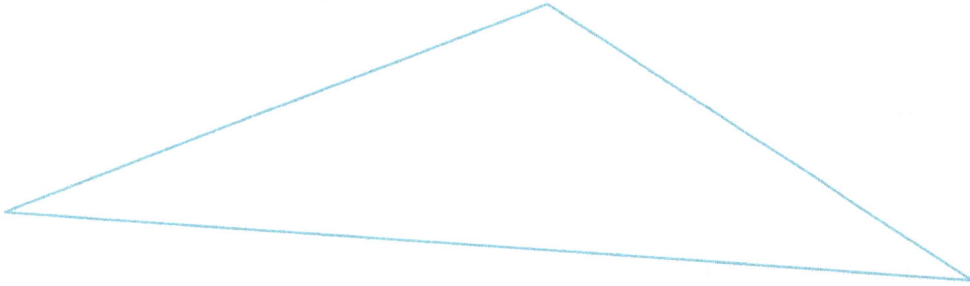

46. Name the quadrilaterals. Use the most descriptive names.

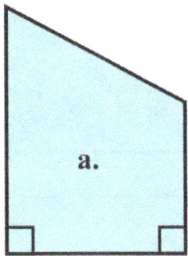

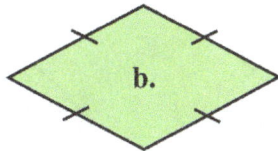

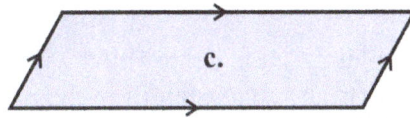

a. _____

b. _____

c. _____

47. Classify each triangle according to its sides and its angles.

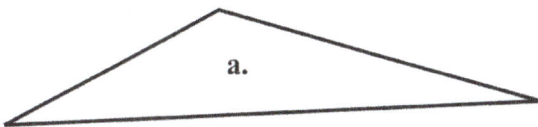

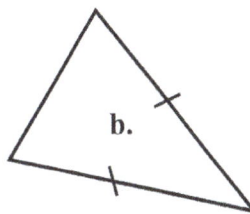

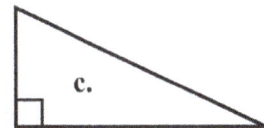

a. _____ and _____

b. _____ and _____

c. _____ and _____

48. Give the definition of a trapezium.

49. Write an "x" if the shape also fulfils the definition of a rectangle or of a parallelogram.

	a rectangle	a parallelogram
a. Every rhombus is also...		
b. Every square is also...		

50. Can an obtuse triangle be isosceles?
    If not, explain why not.
    If yes, sketch an example.

51. A rectangular prism is being filled with little cubes
    (cubic units). The cubes can fit in the prism four
    levels high.

    a. What is the volume of the prism, in cubic units?

    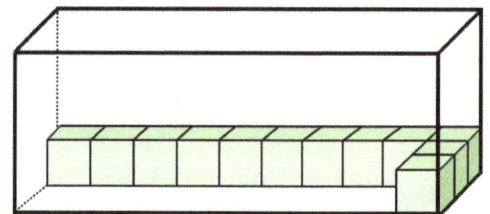

    b. If the edge of each little cube measures 2 cm, what
        is the volume of the prism, in cubic centimetres?

52. This box is a rectangular prism.
    What is the volume of *four* such boxes?

    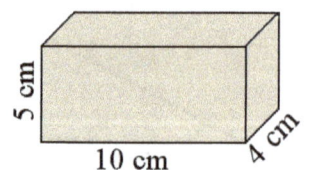
    5 cm
    10 cm
    4 cm

53. Matthew has a rainwater collection tank in his yard that is rectangular, like a box. It is 1.2 m long, 60 cm wide, and 1 m tall.

   a. Find the volume of the tank in cubic <u>metres</u>.

   b. After a rainy night, the tank was about 1/3 full.
   About how many litres of water were in the tank?
   1 cubic metre equals 1000 litres.

Using the Cumulative Revisions

The cumulative revisions practise topics in various chapters of the Math Mammoth complete curriculum, up to the chapter named in the revision. For example, a cumulative revision for chapters 1-6 may include problems matching chapters 1, 2, 3, 4, 5, and 6. The cumulative revision lesson for chapters 1-6 can be used any time after the student has studied the curriculum through chapter 6.

These lessons provide additional practice and revision. The teacher should decide when and if they are used. The student doesn't have to complete all the cumulative revisions. I recommend using at least three of these revisions during the school year. The teacher can also use the revisions as diagnostic tests to find out what topics the student has trouble with.

Math Mammoth complete curriculum also includes an easy worksheet maker, which is the perfect tool to make more problems for children who need more practice. The worksheet maker covers most topics in the curriculum, excluding word problems. Most people find it to be a very helpful addition to the curriculum.

The download version of the curriculum comes with the worksheet maker, and you can also access the worksheet maker online at

https://www.mathmammoth.com/private/Make_extra_worksheets_grade5.htm

Cumulative Revision, Grade 5, Chapters 1-2

1. Place parentheses into these equations to make them true.

   a. $90 + 70 + 80 \times 2 = 390$ **b.** $378 = 6 \times 8 + 13 \times 3$ **c.** $90 \times 4 = 180 - 60 \times 3$

2. Write the numbers in the boxes on the top and on the left side of the large rectangle. Then multiply, and write the area of each part inside it. Lastly calculate the total by adding.

   (It won't matter which number you choose for the horizontal and which for the vertical side.)

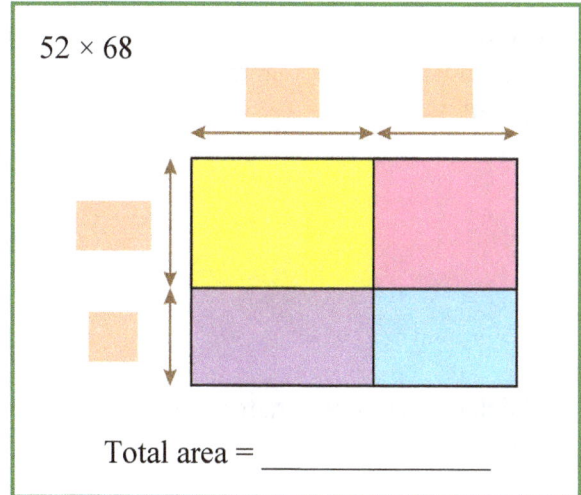

52×68

Total area = _____

3. Divide. Use the space on the left for building a multiplication table of the divisor. Lastly, check.

$2 \times 15 = 30$	**a.** $15\overline{)9450}$	$\times\ 1\ 5$
	b. $14\overline{)4508}$	$\times\ 1\ 4$

4. Solve the word problems.

    a. Jim earned a total of $1920 in four weeks.
       How much did he earn in one week?

    b. Joe entered his sled and dogs in 11 races last year. The
       races were all held on the same 136-km race course. How
       many total kilometres did Joe and his dogs race last year?

5. Which expression(s) match the problem? Also, solve the problem.

Greg bought four flashlights for $9 each, and paid with $50. What was his change?	**(1)** $50 − $9 + $9 + $9 + $9 **(4)** 4 × $9 − $50 **(2)** $50 − ($9 − $9 − $9 − $9) **(5)** $50 − 4 × $9 **(3)** $50 − ($9 + $9 + $9 + $9) **(6)** $50 + 4 − $9

6. First, estimate the answer to the multiplication problem. Then multiply.

a. Estimate: _____	**b.** Estimate: _____	**c.** Estimate: _____
_____	_____	_____
$\begin{array}{r} 1\ 7\ 3 \\ \times\quad 3\ 5 \\ \hline \end{array}$	$\begin{array}{r} 2\ 6\ 9 \\ \times\ 5\ 3\ 7 \\ \hline \end{array}$	$\begin{array}{r} 8\ 9\ 2 \\ \times\ 3\ 4\ 0 \\ \hline \end{array}$

Cumulative Revision, Grade 5, Chapters 1-3

1. Divide. Use the space on the right for building a multiplication table of the divisor. Then check.

$2 \times 21 = 42$	$21 \overline{)\, 8\ 1\ 6\ 9}$	$\times\ 2\ 1$

2. Solve in the right order. You can enclose in a "bubble" or a "cloud," the first operation to be done.

a. $94 + 12 \times 5 \div 2 =$ _____	**b.** $(22 - 9) \times 2 + 58 =$ _____
c. $43 + (55 + 5) \div 5 =$ _____	**d.** $700 - 30 \times (3 + 4) =$ _____

3. Solve mentally.

a. $43 - 17 =$ _____	**b.** $54 - 19 + 12 =$ _____	**c.** $1200 -$ _____ $= 750$
$71 - 43 =$ _____	$85 - 25 + 75 =$ _____	$2000 - 800 -$ _____ $= 600$

4. Write the numbers.

a. 78 billion   38   16 thousand

b. 844 billion   12 million   704

5. Round the numbers as indicated.

number	32 274 302	64 321 973	388 491 562	2 506 811 739
to the nearest 1000				
to the nearest 10 000				
to the nearest 100 000				
to the nearest million				

6. Complete the addition path using mental maths.

43 199 000	add 10 000 →		add a million →	

+ 100 000 ↓

	← add 10 million		← add a thousand	

7. Write an expression to match each written sentence.

a. The product of 5 and 6 is added to 50.	**b.** The difference of 9 and 6 is subtracted from 10.

8. Write a single expression using numbers and operations for the problem. Don't just write the answer!

A teacher bought 21 notebooks for $2 each, 20 rulers for $1.50 each, and chalk for $12. What was the total cost?

9. Add.

a. 521 607 090 + 4 293 991 092	**b.** 77 630 087 + 884 000 299 + 84 926 571

10. Estimate first, using mental maths. Then find the exact answer and the error of your estimation using a calculator.

a. What is your change, if you pay for four shirts that cost $8.90 apiece with $40?	**b.** In a game, Alex bought six jewels for 215 points each, and a shovel for 185 points. How many points did he use in total?
My estimation: _____	My estimation: _____
_____	_____
Exact answer: _____	Exact answer: _____

Cumulative Revision, Grade 5, Chapters 1-4

1. Calculate.

a. $2 \times 10^4 =$ _____

$43 \times 10^5 =$ _____

b. $3090 \times 10^6 =$ _____

$10^8 \times 304 =$ _____

2. Jack earned $125 and his sister earned 4/5 as much. How much did Jack and his sister earn together?

Mark the information in the bar model, and solve.

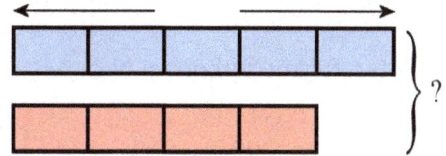

3. In what place is the underlined digit? What is its value?

a. 791 4<u>5</u>6 030

Place: _____

Value: _____

b. 2 09<u>4</u> 806 391

Place: _____

Value: _____

4. Ann is an English teacher. She has 150 students in her English classes this year, and 6/50 of them were not in her classes last year.

   a. How many new students does she have?

   b. Out of the new students, 1/3 have never studied English before. How many of the new students have studied English before?

5. Mark an "x" if the number is divisible by 2, 3, 4, 5, 6, or 9.

Divisible by	2	3	4	5	6	9
692						
3072						

Divisible by	2	3	4	5	6	9
702						
91						

6. Find all the factors of the given numbers. Use the checklist, and keep track of *all* the factors you find.

| a. 35

Check 1 2 3 4 5 6 7 8 9 10

factors: _____ | b. 40

Check 1 2 3 4 5 6 7 8 9 10

factors: _____ |

7. Solve for the unknown N or M.

a. $4 \times M = 200$	b. $M \div 6 = 8$	c. $4500 \div M = 50$
d. $7 \times N = 56\,000$	e. $N \div 30 = 700$	f. $48\,000 \div N = 600$

8. Write an expression to match each written sentence.

| a. The quotient of 350 and x equals 5. | b. The difference of 15 and 6 is added to 8. |

9. Find a number to fit in the box so the equation is true.

| a. $36 = (\boxed{} + 9) \times 3$ | b. $7 \times 7 = 4 \times \boxed{} + 5$ | c. $19 = (84 \div \boxed{}) - 2$ |

10. Round the numbers as indicated.

number	97 302	709 383 121	89 534 890 066
to the nearest 1000			
to the nearest 10 000			
to the nearest 100 000			
to the nearest million			

Cumulative Revision, Grade 5, Chapters 1-5

1. Round the numbers to the nearest unit (one), to the nearest tenth, and to the nearest hundredth.

Round this to the nearest →	unit (one)	tenth	hundredth
4.925			
6.469			

Round this to the nearest →	unit (one)	tenth	hundredth
5.992			
9.809			

2. Add using mental maths.

a. $0.3 + 0.07 =$ _____	**b.** $0.19 + 0.002 =$ _____	**c.** $0.028 + 0.3 =$ _____
d. $1.05 + 0.4 =$ _____	**e.** $0.49 + 0.56 =$ _____	**f.** $0.006 + 0.5 =$ _____

3. Jerry bought three packets of AA batteries and six packets of AAA batteries.
   The total was $100.50. One packet of the AA batteries cost $7.90.
   What does one packet of AAA batteries cost?

4. Find the prime factorisation of the numbers. If the number is prime, write it as 1 times the number.

a. 45 /\	**b.** 23 /\	**c.** 64 /\

5. Write in expanded form.

   a. 0.908

   b. 543.2

6. These decimal divisions are not even. Round the answers to the nearest hundredth.

a. $3.377 \div 3$	**b.** $22.91 \div 11$	**c.** $62.6 \div 7$
$\overline{)}$	$\overline{)}$	$\overline{)}$

7. Divide in two ways: first by indicating a remainder, then by long division. Give your answers to two decimal digits.

a. $31 \div 6 =$ _____ R _____	**b.** $43 \div 4 =$ _____ R _____
$\overline{)}$ Check:	$\overline{)}$ Check:

8. Solve for x.

a.
```
├──── x ────┤
[ 14 ][  ][  ][  ]
```

b.
```
├───── 1500 ─────┤
[ x ][  ][  ][  ][  ]
```

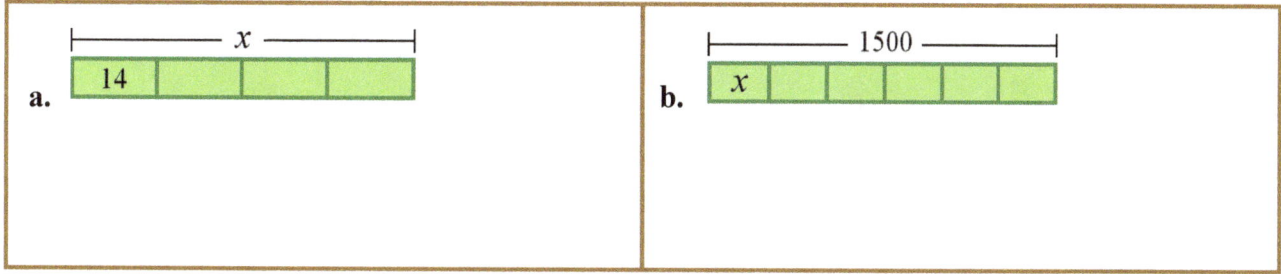

9. Calculate the products mentally.

a. $600 \times 20\,000$	b. $8 \times 30 \times 100\,000$	c. $700\,000 \times 8000$

10. Helen has 120 marbles and Julie has 2/5 as many.
    How many more marbles does Helen have than Julie?

11. How many seconds are there in an hour?

    How many seconds are there in a day?

12. A hotel maintains two jogging paths in the woods. The shorter one
    is 1.2 km long and the other is four times as long. If you jog both
    paths, then what is the total distance you have jogged?

Cumulative Revision, Grade 5, Chapters 1-6

1. First *estimate* the answer to each multiplication. Then multiply to find out the exact answer.

a. 290×277	**b.** 525×416	**c.** 897×186
Estimate:	Estimate:	Estimate:
X _____	X _____	X _____

2. Angie and Rebecca split their total earnings of
   $100 so that Angie got $10 more than Rebecca.
   How much did each one get?

3. Add and subtract in your head.

a. $0.08 + 2 =$ _____	**b.** $0.02 + 0.6 =$ _____	**c.** $0.005 + 0.1 =$ _____
d. $2 - 0.09 =$ _____	**e.** $0.5 - 0.06 =$ _____	**f.** $0.06 - 0.005 =$ _____

4. Write using exponents, and solve.

a. $10 \times 10 \times 10 \times 10 =$	**c.** $2 \times 2 \times 2 \times 2 \times 2 =$
b. $100 \times 100 \times 100 =$	**d.** six cubed $=$

5. Calculate the products mentally.

a. 13×10^9	**b.** 7302×10^5

6. Solve the equations.

a. $y - 0.57 = 1.1$	**b.** $7.319 + z = 9$

7. Calculate the average (the mean) of the data set.
   Do not use a calculator.

   21, 19, 25, 22, 13, 15, 24, 12, 11

8. A group of 37 medical students travelled through ten
   states to view new technology in some progressive
   hospitals. They had to share the expense of $99 000 for
   the trip. What was each student's share of the expenses?
   Round your answer to the nearest dollar.

9. Make a line graph of this data for the Oak Bend Hospital. Note that you need to choose the scaling
   for the axes.

Year	Babies Born
1950	225
1960	340
1970	460
1980	525
1990	580
2000	520
2010	490

10. Add 406 292 399 + 35 290 911 + 46 329
without using a calculator.

11. Which expression would solve the problem? Also, find the answer.

A book that is 5 cm thick is lying in a box that is 17 cm high.
How many books that are 2 cm thick could you stack in the box
on top of that book?

$17 \times 5 + 2$

$(17 - 5) \div 2$

$(17 - 2) \div 5$

$17 + 5 + 2$

$(17 \div 5) + 2$

$(17 \div 2) + 5$

12. Find a number to fit in the box so the equation is true.

a. $34 = (\boxed{} + 11) \times 2$	**b.** $3 \times 8 + 5 \times 6 = \boxed{} - 10$	**c.** $8 + 5 = (50 - \boxed{}) \div 2$
d. $81 = 9 \times (11 - \boxed{})$	**e.** $5 \times \boxed{} = 120 - 20 \times 3$	**f.** $\boxed{} \div 2 = (20 + 5) \times 3$

13. Plot the points from the "number rules" in the coordinate grid. Note that the coordinate grid is scaled differently.

Rule for x-values: start at 0, and add 10 each time.

Rule for y-values: start at 2, and add 1 each time.

x	0	10	20	30	40	50	60	70	80	90	100	110
y	2	3	4				8					

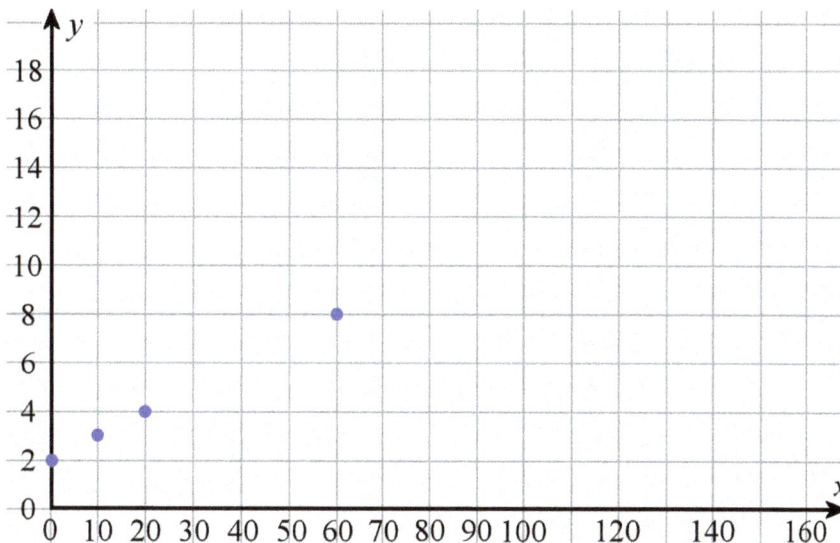

Cumulative Revision, Grade 5, Chapters 1-7

1. Solve in the right order!

a.	b.	c.
$13 \times 4 + 18 =$ _____	$(2 + 60 \div 4) \times 3 =$ _____	$10 \times (9 + 18) \div 3 =$ _____
$4 + 8 \div 8 =$ _____	$2 + 30 \times (7 + 8) =$ _____	$5 \times (200 - 190 + 40) =$ _____

2. Joe bought 100 apples for $0.23 each. He divided them equally into ten small bags.

  a. What was the total cost for 100 apples?

  b. What was the value of each small bag of apples?

3. Find the missing factor.

a. $10 \times$ _____ $= 4.0$	**c.** _____ $\times 0.11 = 3.3$	**e.** $2 \times$ _____ $\times 1.2 = 48$
b. $5 \times$ _____ $= 6.0$	**d.** _____ $\times 0.3 = 0.06$	**f.** $3 \times$ _____ $\times 0.5 = 6$

4. Divide. Mental maths will work!

a. $0.8 \div 2 =$ _____	**b.** $0.36 \div 6 =$ _____	**c.** $0.25 \div 0.05 =$ _____
d. $0.16 \div 4 =$ _____	**e.** $0.54 \div 0.06 =$ _____	**f.** $1 \div 0.05 =$ _____

5. **a.** Design a scaling for the axes so that the point (35, 40) will fit on the grid.

  b. Plot the points (35, 40), (35, 25), (20, 15) and (20, 30) on the grid. Join the points in order with line segments.

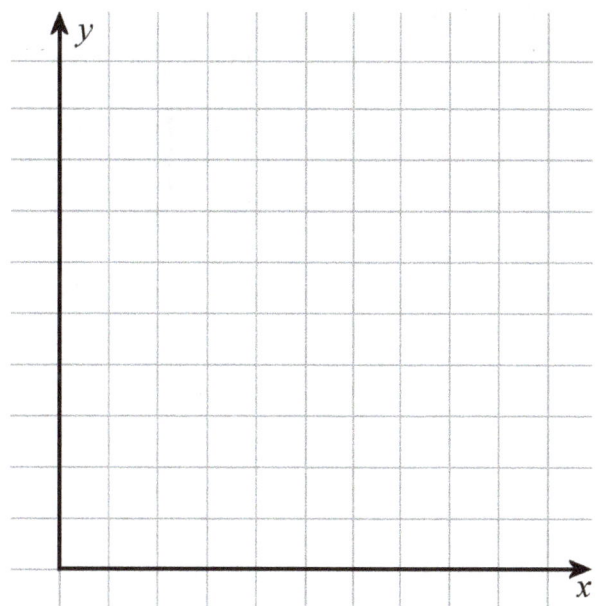

6. Divide.

a. $0.927 \div 0.3$	b. $0.646 \div 0.08$
Check:	Check:
$\overline{)}$	$\overline{)}$

7. Ashley bought a one-litre carton of milk, and used 1/4 of it for baking. How many *millilitres* of milk are left?

8. Ava is 142 cm tall and Eva is 0.66 metres tall. Who is taller? How many centimetres taller?

9. Convert the measuring units.

a.		b.		c.	
0.5 m = _____ cm		4.2 L = _____ mL		800 g = _____ kg	
0.06 m = _____ cm		400 mL = _____ L		4550 m = _____ km	
2.2 km = _____ m		5400 g = _____ kg		2.88 kg = _____ g	

10. While jogging, Rebecca saw a big snake on the path 250 m before the end of the 2.4-km jogging track. She got so scared that she turned back on the track and jogged back to the beginning of the track. Find the total distance that she jogged on the track.

11. First, estimate the answer. Then, multiply. Do not forget the decimal points.

a. 2.09×11.5	**b.** 73×2.14	**c.** 7.1×3.02
Estimate: _____	Estimate: _____	Estimate: _____
X _____	X _____	X _____

12. Alex bought seven packets of cucumber seeds for a total of $13.23. He also bought seven flowering plants that were originally $3.20 each but the price was reduced by 4/10.

   a. What did one packet of seeds cost?

   b. How much did one flowering plant cost?

   c. What was the total cost?

12. Shelly is going to buy four kilograms of oranges for $3.10 a kilogram, and three kilograms of bananas for $2.95 a kilogram.

   a. Estimate the total cost to the nearest dollar.

   b. The clerk informs Shelly that she will get 1/5 off of the total cost for being a loyal customer. Now calculate what Shelly pays. Use the actual total in your calculation, not the rounded total.

Cumulative Revision, Grade 5, Chapters 1-8

1. Solve by multiplying in columns.

a. 21.7×3.9	**b.** 0.52×0.8	**c.** 141×5.22
Estimate: _____	Estimate: _____	Estimate: _____
X _____	X _____	X _____

2. A drinking glass measures 3/10 of a litre.
   How many glasses can you fill from a 3-litre pitcher full of water?

3. Juan is mailing 36 books that weigh 170 g each. What is
   the total weight of the books, in kilograms and grams?

4. Find the prime factorisation of the numbers.

a. 55 /\	**b.** 75 /\	**c.** 48 /\

5. Compare the fractions.

a. $\dfrac{2}{3}$ ☐ $\dfrac{5}{8}$ b. $\dfrac{1}{4}$ ☐ $\dfrac{4}{9}$ c. $\dfrac{5}{6}$ ☐ $\dfrac{5}{7}$ d. $\dfrac{6}{8}$ ☐ $\dfrac{3}{4}$

6. Mum is buying a thermometer, and the store has two kinds.
   The more expensive one costs $28.40, and the cheaper just 3/4 as much.
   How much more does the more expensive thermometer cost than the cheaper one?

7. Cassandra sells orange juice. Below you see the total amounts of juice, in litres, that she sold on
   11 different days. Make a line plot.

   3   4 ¼   3 ½   3 ½   3 ¾   3 ¾   3 ¾   3 ¼   4 ¾   3 ¼   2 ¾

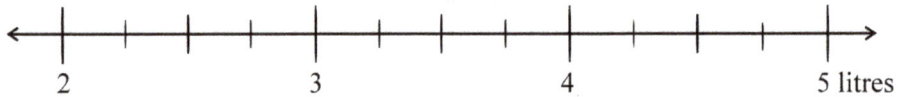

8. Add and subtract.

a. $6\dfrac{6}{11} - 3\dfrac{2}{5}$	b. $6\dfrac{6}{7} + 1\dfrac{1}{2}$
c. $7\dfrac{9}{10} - 1\dfrac{1}{4}$	d. $3\dfrac{2}{5} + 2\dfrac{5}{6}$

9. Multiply and divide.

a. $10 \times 0.07 =$ _____	**b.** $100 \times 0.63 =$ _____	**c.** $10^5 \times 0.029 =$ _____
d. $0.8 \div 10 =$ _____	**e.** $4.5 \div 100 =$ _____	**f.** $76 \div 10^3 =$ _____

10. **a.** Fill in the table how much weight Greg gained during each year.

    b. When did he gain weight fastest: from age 0 to 1 year, or from age 4 to 5, or from 13 to 14?

    c. What year (from what age to what age) did he gain the least weight?

AGE (yrs)	WEIGHT (kg)	Weight gain from previous year
0	3.3 kg	N/A
1	10.2 kg	6.9 kg
2	12.3 kg	2.1 kg
3	14.6 kg	2.3 kg
4	16.7 kg	2.1 kg
5	18.7 kg	2.0 kg
6	20.7 kg	
7	22.9 kg	
8	25.3 kg	
9	28.1 kg	

AGE (yrs)	WEIGHT (kg)	Weight gain from previous year
10	31.4 kg	
11	32.4 kg	
12	37.0 kg	
13	40.9 kg	
14	47.0 kg	
15	52.6 kg	
16	58.0 kg	
17	62.7 kg	
18	65.0 kg	2.3 kg

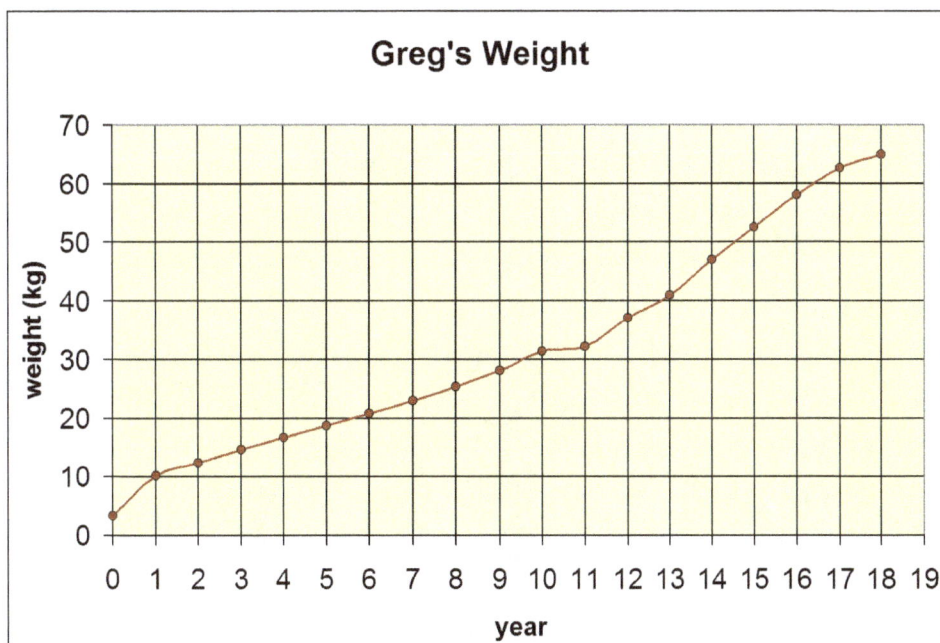

Greg's Weight

Cumulative Revision, Grade 5, Chapters 1-9

1. Shade a rectangle inside the square so that its area can be found by the fraction multiplication.

a. $\dfrac{3}{4}$ m $\times$ $\dfrac{1}{2}$ m = ——— m^2 b. $\dfrac{2}{3}$ cm $\times$ $\dfrac{5}{6}$ cm = ——— cm^2

2. Classify each triangle by its angles, and by its sides.

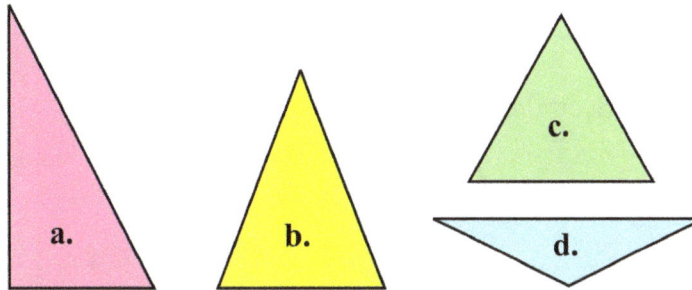

Triangle	Classification by the angles	Classification by the sides
a.		
b.		
c.		
d.		

3. The volume of a box is 108 cubic centimetres. Find its width if it is 12 cm deep and 3 cm tall.

4. Name these quadrilaterals.

a. _____

b. _____

c. _____

d. _____

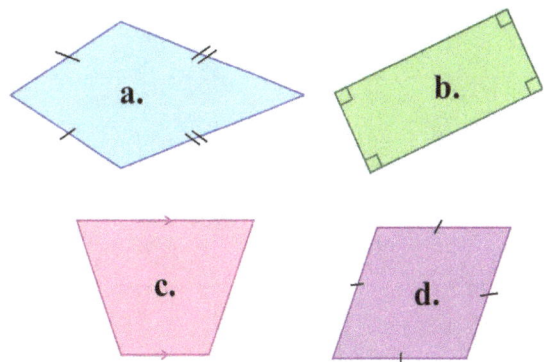

5. **a.** Is $\dfrac{11}{13} \times \dfrac{3}{4}$ more or less than $\dfrac{3}{4}$?

b. Explain in your own words why that is so.

c. Find a value for n so that $\dfrac{n}{10} \times \dfrac{1}{3}$ is more than $\dfrac{1}{3}$.

6. Scott is a plumber, and each day he has to drive around town to the clients' homes. The following numbers show how many kilometres Scott drove on ten different workdays:

128 68 73 163 93 102 68 85 90 45

a. Find the average distance Scott drove per day.

b. Based on the average, calculate *approximately* how many kilometres Scott would drive at work in a year's time. Assume that he works 40 weeks a year, 5 days a week.

7. Multiply. Give your answers in the lowest terms (simplified) and as a mixed number, if possible.

a. $\dfrac{6}{8} \times \dfrac{2}{9}$	**b.** $\dfrac{9}{11} \times 2\dfrac{1}{3}$

8. Divide.

a. $1 \div \dfrac{1}{3}$	**b.** $5 \div \dfrac{1}{4}$	**c.** $\dfrac{1}{3} \div 3$
d. $\dfrac{1}{6} \div 5$	**e.** $7 \div 3$	**f.** $\dfrac{9}{10} \div 3$

9. Solve.

a. A room that is 7 metres long is divided into four equal partitions.
   How long is each partition?

b. Find two-thirds of 3/8 of a tonne load of gravel.

c. A farmer divides a 50-kg sack of feed equally into eight parts.
   How many kilograms is each part?

10. A pan full of brownies was cut into 24 pieces. After a day,
    2/3 of it was left. Then the family ate 3/4 of the
    remaining brownies. How many pieces are left now?

11. Check your skills in all four operations of decimal arithmetic, and calculate:

a. $4.59 + 0.109 + 350.8$ b. $2.04 - 1.297$

c. 6.5×13.8 d. $4.6 \div 0.008$

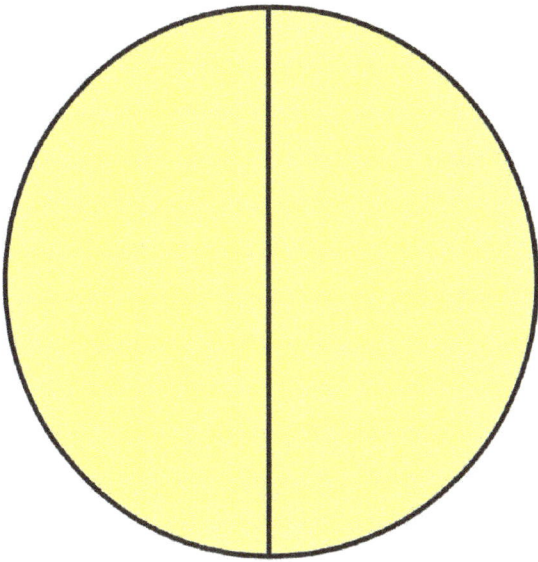 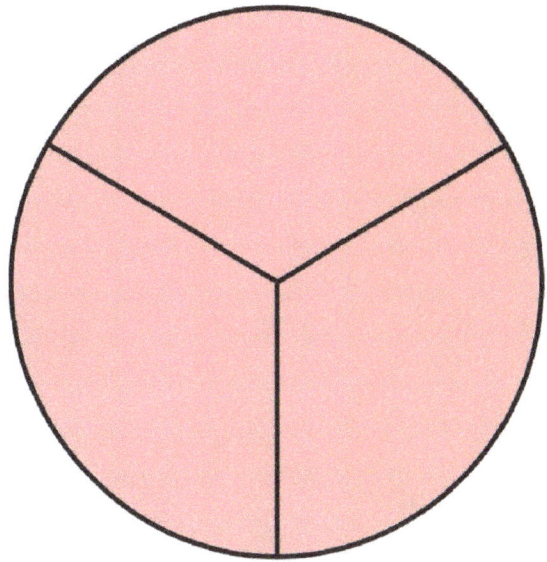

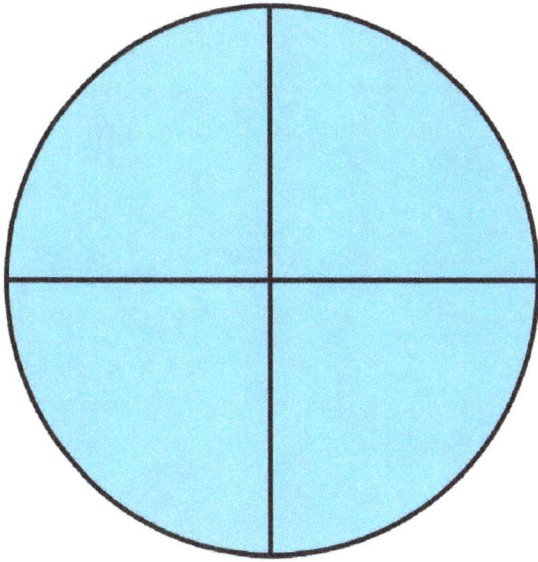 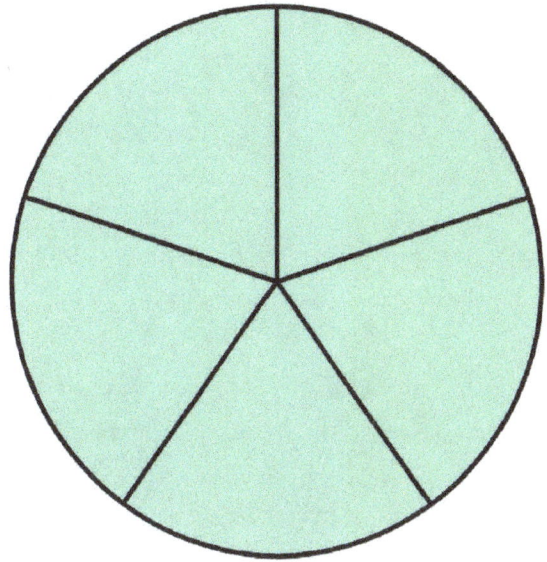

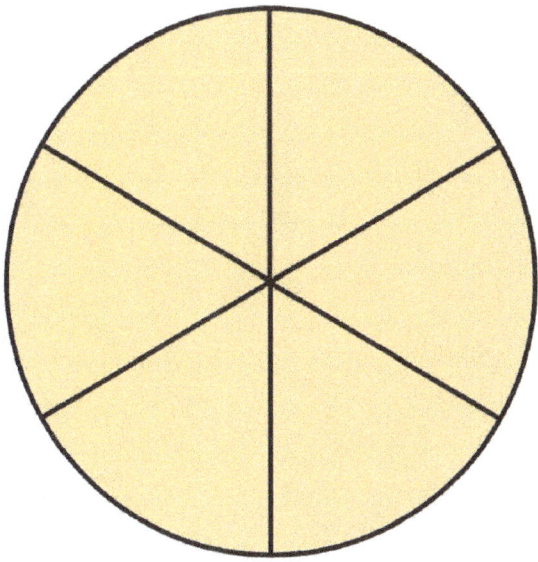 

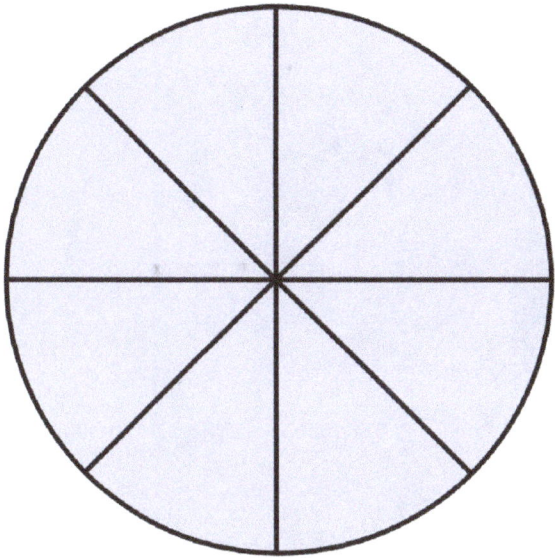
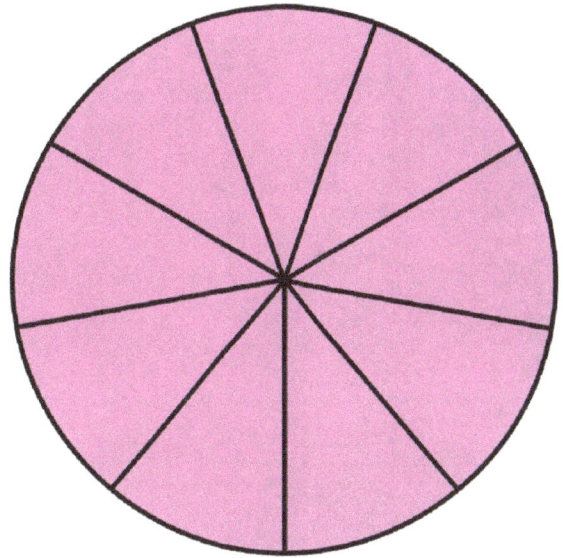
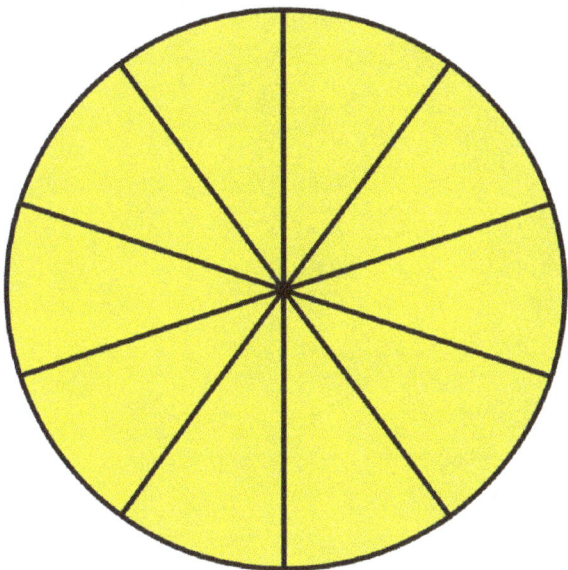
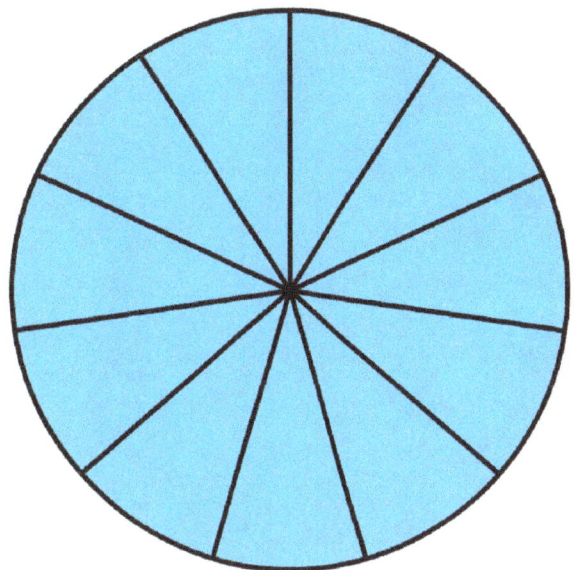
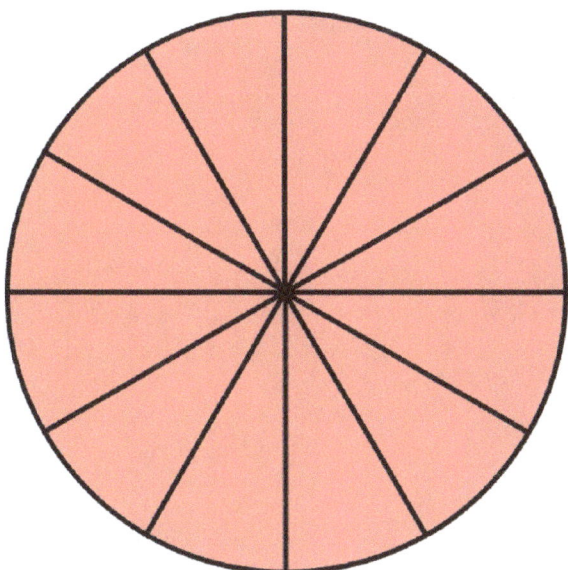
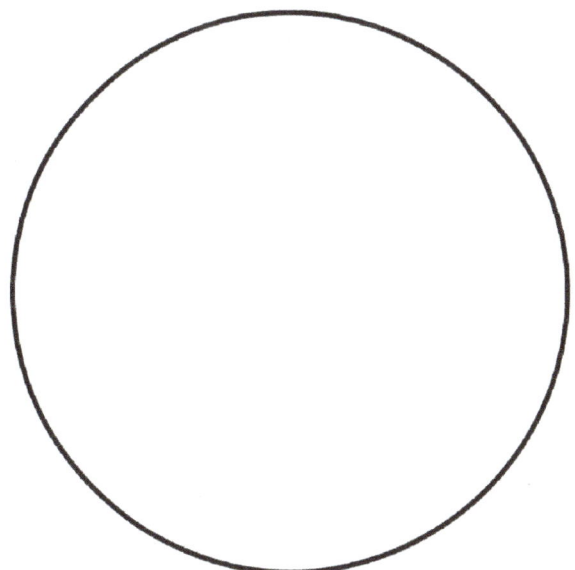

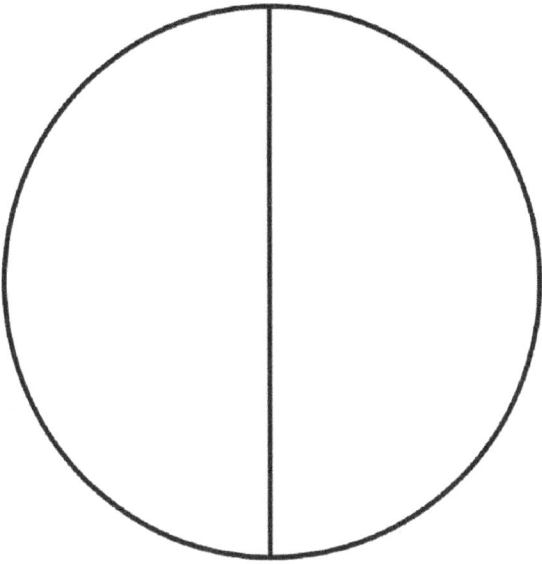
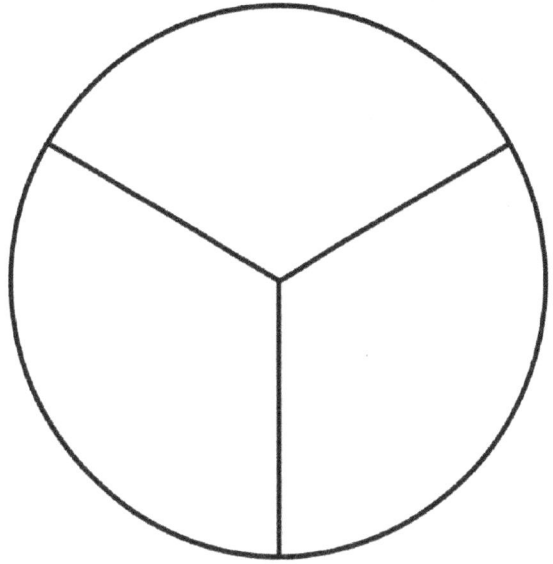
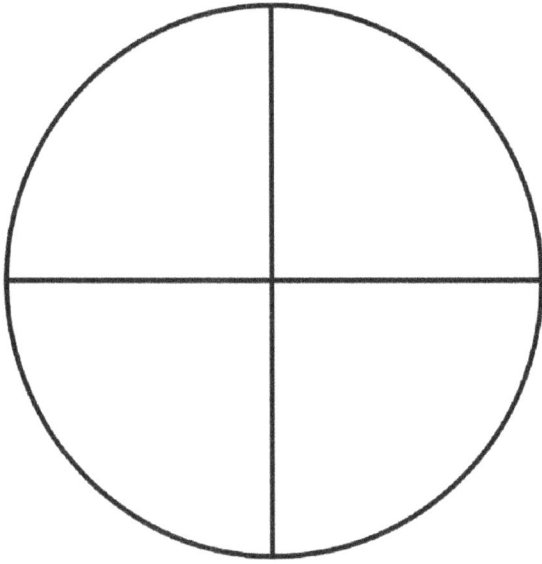
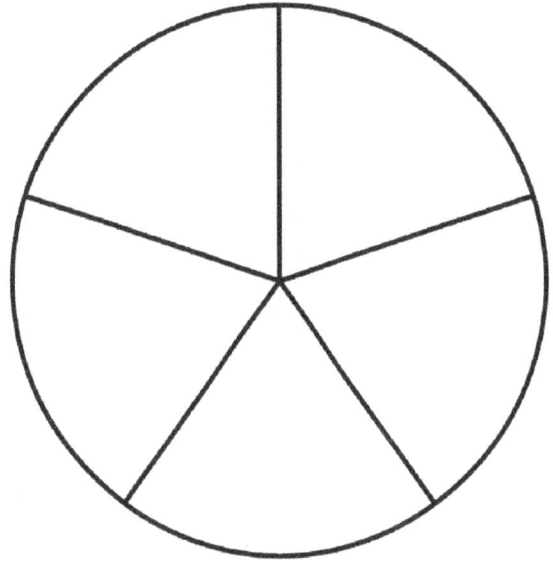
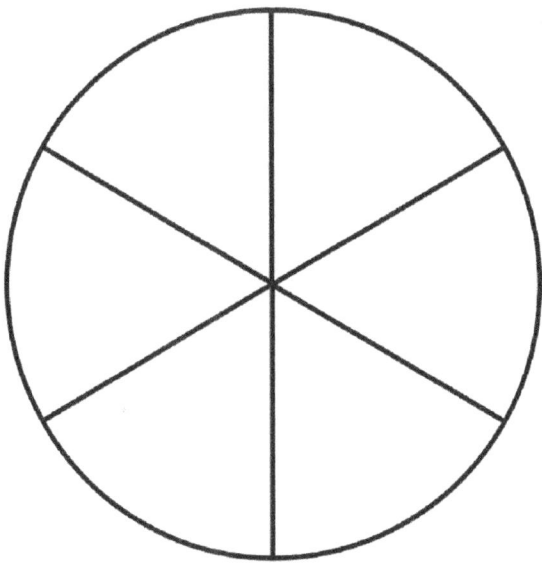
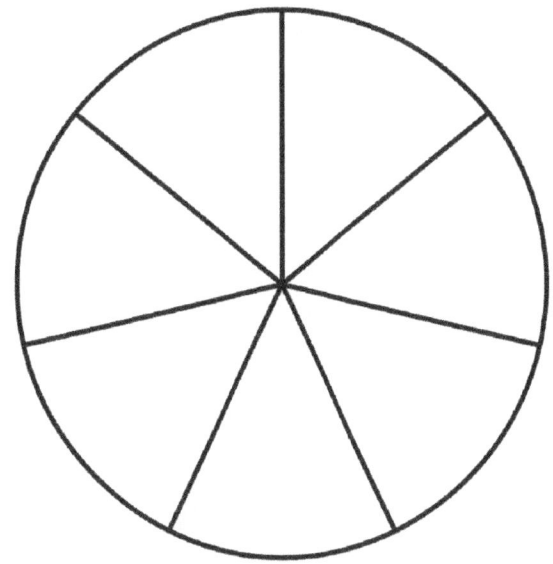

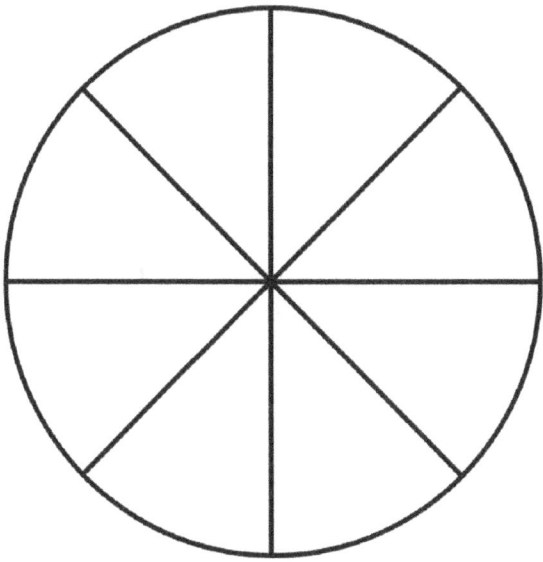
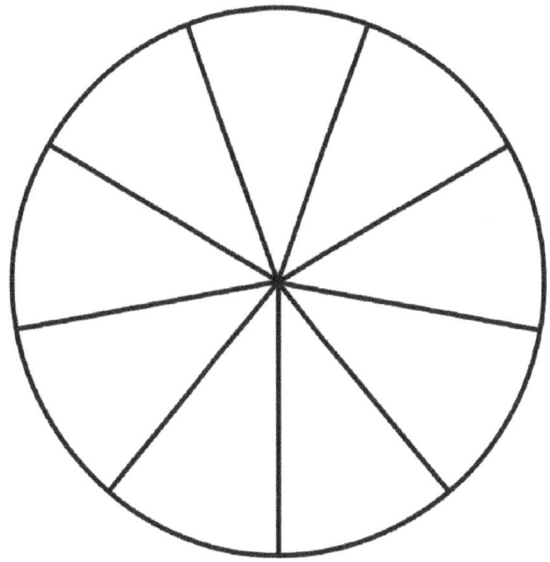
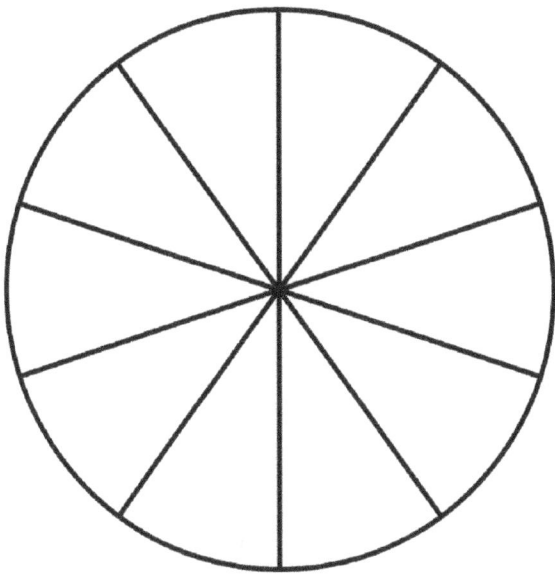
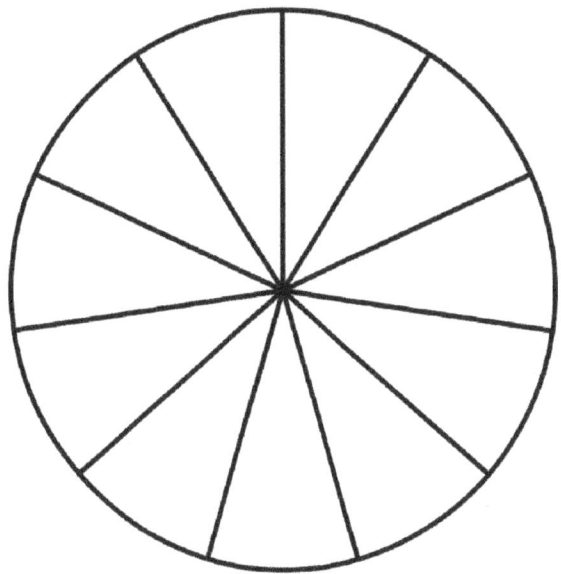
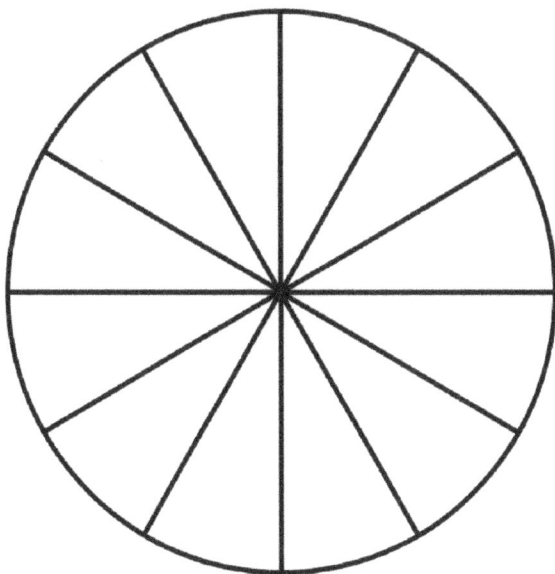
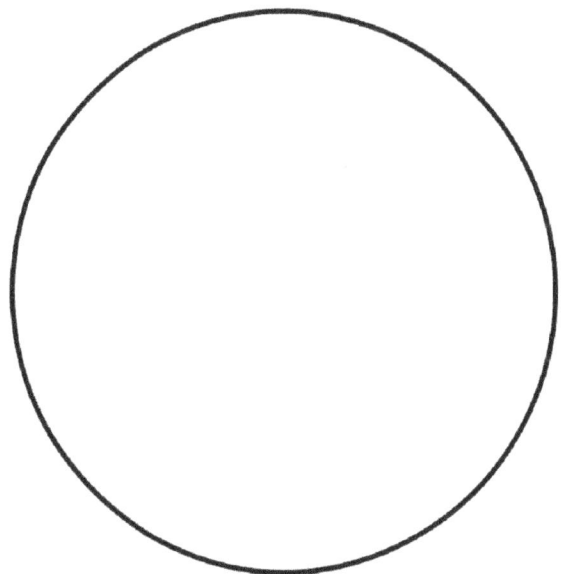

www.ingramcontent.com/pod-product-compliance
Lightning Source LLC
Chambersburg PA
CBHW080252200326
41519CB00023B/6964